D1431883

CONTESTED MEDICINE

CONTESTED MEDICINE

CANCER RESEARCH AND THE MILITARY

Gerald Kutcher

THE UNIVERSITY OF CHICAGO PRESS *Chicago & London*

GERALD KUTCHER is Dean's Professor of the History of Medicine at
SUNY Binghamton.

The University of Chicago Press, Chicago 60637
The University of Chicago Press, Ltd., London
© 2009 by The University of Chicago
All rights reserved. Published 2009
Printed in the United States of America

18 17 16 15 14 13 12 11 10 09   1 2 3 4 5

ISBN-13: 978-0-226-46531-9
ISBN-10: 0-226-46531-4

Library of Congress Cataloging-in-Publication Data
Kutcher, Gerald.
    Contested medicine : cancer research and the military / Gerald
Kutcher.
        p.   cm.
    Includes bibliographical references and index.
    ISBN-13: 978-0-226-46531-9 (hardcover : alk. paper)
        ISBN-10: 0-226-46531-4 (hardcover : alk. paper)   1. Cancer—
Research—United States—History—20th century.   2. Human
experimentation in medicine—United States—History—20th
century.   3. Clinical trials—Moral and ethical aspects—United
States—History—20th century.   4. Radiation—
Physiological effect—Research—United States—History—20th
century.   5. Medicine, Military—Research—United States—
History—20th century.   I. Title.
    [DNLM: 1. United States.   2. Human Experimentation—
ethics.   3. Neoplasms—radiotherapy.   4. Clinical Trials as
Topic—ethics.   5. Military Medicine—history.   6. Whole-
Body Irradiation—history.   QZ 269 K975C   2009]
    RC267.K88   2009
    362.196'9940072—dc22   2008024075

♾ The paper used in this publication meets the minimum requirements
of the American National Standard for Information Sciences—
Permanence of Paper for Printed Library Materials, ANSI Z39.48-1992.

To the memory of

Emmanuel van der Schueren (1943–98),

a prince among men

CONTENTS

ACKNOWLEDGMENTS

This book on human experimentation reflects, in part, the perspectives of a practitioner who had a long career in radiation medicine. It is also a story written by the educated outsider. But to be able to write from this latter point of view proved especially demanding—it meant that I had to learn to see the world of experimentation anew, and that entailed a period of (nearly) full-time study and writing shielded from the daily demands of hospital life. To make such a major adjustment in my life was not easy, and I have a number of individuals to thank for their support along the way. I want to especially acknowledge Simon Schaffer, who introduced me to new ways of understanding experimental life and engaged me in fascinating and illuminating discussions on every aspect of this work. Special thanks go to Chris Clarke and Allan Brandt for their encouragement and advice as I contemplated taking on this project. John Forrester and Samuel Hellman provided helpful and intelligent readings of some of the early essays. Harry Bartelink offered a careful and positive reading of an earlier version of this text. It is a pleasure to recognize Nick Hopwood for his insightful suggestions on some of my essays and for his advice and encouragement and Soraya de Chadarevian for her astute comments on much of this book. Marjorie Ciarlante at National Archives II provided me with what seemed like an endless number of boxes of documents from the ACHRE files; the staff at the Countway Library at Harvard University helped with the Henry Beecher papers. I want to thank Dean Peter Mileur and Harpur College, Binghamton University, for awarding me a research leave for the spring semester of 2007, during which time I completed this manuscript. I also want to acknowledge the fellows of Clare Hall, University of Cambridge, for electing me a visiting fellow for the same period and providing an ideal work environment. Special thanks go to Christie Henry, my editor at the University of Chicago Press, for her support of this project. Finally, I want to acknowledge Marilynn Desmond, who encouraged me to take the leap. Marilynn, who is a marvelous reader, went through every version of this text and pointed out myriad examples of poor writing and countless contradictions. But she did something more important. She encouraged me—better still, inspired me—to never settle for the mundane, to search for new ways to interpret the archival materials, to write yet one more version, and, especially, to constantly aspire to reach higher.

I dedicate this book to the memory of Emmanuel van der Schueren, who died on March 3, 1998, and was then director of the University Hospital of Leuven, Belgium, and previously chief of oncology. Without his support, I could not have pursued this work. When I was struggling to make a decision whether to take on this project and leave my position at Memorial Sloan-Kettering, Emmanuel guaranteed me at least three years of funding as a part-time consultant at the University Hospital of Leuven. This arrangement would provide the financial support to enable me to carry out my goals. I vividly remember a lunch we had in New York shortly before I made my decision. We shook hands over the proposed arrangement, and I told Emmanuel that the handshake was all I required. He prophetically shook his head and said that we should put it in writing to protect me: "You never know, I could be run over by a bus." Within days of my arrival in the United Kingdom in January 1998 to begin my work on this project, I learned that Emmanuel had just been diagnosed with pancreatic cancer. I understood that the prognosis was dire, and I was deeply saddened that I would shortly lose a dear friend and colleague. I also despaired that my project would be ended even before it had begun since I had quit my job, had no source of support, and was sure that nothing about my arrangement with Emmanuel had been put in writing. I saw Emmanuel twice in his hospital room over the coming months but did not ask him about the state of the funds. How could I? Some time later, I learned that, three days before he died, Emmanuel had transferred the three years of funding to Dominique Huyskens, the chief of physics, and instructed him to make sure it was used solely for my support; otherwise, he was warned, the department would divide it up. I want to thank Dominique for faithfully carrying out Emmanuel's wishes, at, I suspect, some personal cost. Recounting this story is more than a way of thanking Emmanuel; it is also a paean to his character.

INTRODUCTION

These human experiments could be among the most egregious that have been brought to light yet. · Statement before Congress of Representative John Bryant

The...studies...were based on a reasonable hypothesis, were conducted and reported in the scientific literature in keeping with clinical investigations of that period, and seem to have used the accepted standards of informed consent for that period. · Statement before Congress of Dr. James D. Cox

The physician and scientist move in a no-man's-land into which at one point the sovereign alone could penetrate. · Giorgio Agamben, *Homo Sacer*

For over ten years, starting in 1960, patients at the University of Cincinnati College of Medicine were treated for advanced cancers; at the same time, they were experimental subjects who were used to answer questions for the U.S. military on the effects of total-body irradiation (TBI). The combined military medical and cancer therapy studies at Cincinnati have been subjected to numerous investigations, yet they have eluded satisfactory resolution and remain controversial to this day. Some have argued that they were among the most egregious experiments of the cold war period. Others have claimed that they were carried out in keeping with the standards of medical experimentation on human subjects. These two positions, which are typical of the controversy surrounding the case, appear to hopelessly contradict one another. But are they entirely in conflict, or does clinical research by its nature encompass both the normal and the pathological? Is it not possible for research typically considered normal to at times share elements of the egregious and for research typically considered egregious to contain elements of the normal? If so, we should be able to use the experiments in Cincinnati to interrogate both the egregious and the normal facets of research with humans in the post–World War II period. Indeed, during this period, the disturbing character of human studies—for example, the inherent conflict between research and therapy—could no longer be ignored.

Following the Second World War, the medical research enterprise and cancer research, in particular, went through massive changes in scale and scope. This development, however, built on a number of modern practices that were already in place prior to the war. For example, by the 1950s, the voluntary hospital had already become the primary arena for performing clinical research as well as the site for most medical care.[1] Likewise, the large-scale clinical investigations so characteristic of the postwar period had prewar precursors in a number of cooperative studies.[2] Following the war, a huge infusion of money led to a vast increase in the number of clinical studies and in the scale of those programs. This situation put more power in the hands of clinical researchers and, thereby, transformed the character of medical care. In addition, the size of clinical research programs led to modifications in research practices, which, in turn, influenced clinical conduct and, eventually, led to a change in the ethical regulation of medical research.[3]

Cancer therapy and research also went through major changes in the postwar period. In the late 1940s, the National Cancer Institute (NCI) was little more than a moribund appendage of the Public Health Service. Within two decades, however, it had been transformed by an emergent group of chemotherapists into a powerful enterprise that dominated cancer therapy, and its importance only grew further following President Nixon's "War on Cancer." In his January 1971 State of the Union Address, Nixon declared a total commitment to conquering the dread disease of cancer. By the end of the 1970s, the congressional bill that followed his address led to appropriations of nearly $1 billion, and it also provided the NCI with special status among the various programs at the National Institutes of Health (NIH).[4] Alongside these developments, cancer research techniques went through major changes. Although the randomized clinical trial did not make its first appearance in the United States until after the war, by the late 1950s large-scale multicenter trials were perceived to be the only reliable means of answering clinical questions in cancer therapy. In addition, the NCI's cancer drug development program, under strong pressure from the pharmaceutical industry, changed dramatically from the cottage industry that is was just before the war into a large-scale, highly engineered enterprise testing millions of drugs per year.[5]

Radiation therapy went through a major transformation in the postwar period as well, aided by the wartime development of microwave electronics and nuclear energy.[6] These new technologies were used to produce the high-energy radiation machines that made it possible to use aggressive therapy on diseased organs in any part of the body. The growth of large-scale programs and the continuing introduction of expensive technologies were typical not just of postwar medicine; a similar perspective pervaded efforts by the state in its mobilization for military conflict. Not only did Nixon's War on Cancer bring the metaphor of a worldwide conflict to the medical arena, but it also symbolized an underlying belief in the importance of strong science for the power of the state and for the ultimate triumph of America in a time of (cold) war.[7]

The postwar period of medical expansion also saw the establishment of new cancer disciplines. In the 1950s, the cancer chemotherapist was an internist who stole time from clinical duties to practice with new cancer drugs, and the radiation therapist was often a trained general radiologist doing part-time duty in radiation therapy. By the 1970s, medical oncology and radiation therapy had become independent medical disciplines, and, along with surgery, these specialties formed what has become the contemporary triad of cancer therapy. By the 1980s, the cancer community was able to claim some important victories in its War on Cancer. For example, Hodgkin disease and childhood leukemia, previously fatal, showed impressive long-term survival rates with new regimens of radiation therapy, multiagent chemotherapy, and bone marrow transplantation.[8] And, by the late 1980s, radical mastectomy to treat primary breast cancer had begun to give way to the less mutilating combination of localized surgery and radiation.[9]

But there was a disquieting underside to this huge medical research enterprise. By the late 1970s, the cancer therapy community came under attack for not keeping the promises of the War on Cancer.[10] Many saw the new medicine in a disquieting light, as a nightmare world of distanced physicians and cold and impersonal hospitals.[11] In addition, there was another problem in the march of postwar research: it carried a cost in human suffering that, at the start, had not been sufficiently appreciated by the public or, for that matter, by many physicians and public administrators. By the mid-1960s, public trust in medical research had become deeply eroded by a series of highly publicized scandals, and regulators and some medical researchers feared that the research enterprise itself might be in jeopardy. A system of peer review was established in order to transfer ethical judgments in research from the probity of the individual investigator to medical/government institutions.[12] This shift in the location of trust was only the first in a series of government initiatives that diminished the prerogatives and authority of physicians. In addition, with the emergence of the bioethics movement in the 1970s, ethicists began to argue that the basic relationship between patients and physicians was a contractual one underwritten by informed consent, a stance that affirmed and protected not so much the welfare of patients as their rights as autonomous subjects.[13] Patients consequently began to increasingly rely on making their own decisions regarding the therapeutic advantage of a proposed experimental regimen, thereby placing less trust in physicians and medical/government oversight bodies.

The experimental program at the University of Cincinnati highlights some of the issues regarding research and ethical conduct as they emerged during the postwar period. Beginning in 1960, Dr. Eugene Saenger (1917–2007), a radiologist at the University of Cincinnati, delivered TBI to patients with advanced cancers. Saenger, who was under contract with the Department of Defense (DOD), used the patients as proxies for soldiers in order to study radiation effects from nuclear attack. For example, he used chemical measurements of urine samples from the irradiated patients to try to develop a way of indicating the amount of radiation a soldier had received. He claimed that the TBI treatments he gave were intended to treat advanced cancer and that the military component of the study was only a secondary goal, a data-gathering operation that had no effect on clinical decisions. Nevertheless, there were a series of contentious reviews of Saenger's program that began in 1966, shortly after local peer review was instituted in Cincinnati. In spite of the concerns of some of Saenger's peers, his program continued until public disclosures of unethical practices in the fall of 1971 brought the experiments to a temporary halt. The public allegations set in motion a number of investigations by internal and external committees. With one exception, those committees' reports were only mildly critical and generally supportive of the TBI research. Although, by early 1972, it appeared that the TBI program had been vindicated, public outcry that Saenger had secretly negotiated a contract renewal forced the president of the university to terminate contractual arrangements with the DOD. In the mid-1990s, Saenger's program came to the public's attention once again following a rash of press reports about unethical cold war experimentation; at that point, President Bill Clinton created the so-called Advisory Committee on Human Radiation Experiments (ACHRE). As part of its charge, ACHRE looked in great detail at the Saenger case, but, in its *Final Report*, the committee could not reach a definitive ethical judgment on Saenger's research program. Even the legal suits brought by the family members of some of the patients against Saenger and his coinvestigators ended in a 1999 settlement that did not fully address the complex and troubling issues surrounding the case.

Saenger's program was by no means the only case of this type of human experimentation. For example, the two leading cancer centers in the United States, M. D. Anderson Hospital and Tumor Institute in Houston and Memorial Sloan-Kettering Cancer Center in New York, both carried out similar studies.[14] How was it possible for Saenger and other investigators to use cancer patients to answer military questions? Were they simply malign physicians, or were there other forces that encouraged and abetted their efforts? If we look at the cold war environment, we can find a number of reasons why it provided conditions that were favorable to such studies. First, following the Soviet Union's test of an atomic bomb in 1949, politicians, military and civilian planners, and the public began to fear a nuclear attack and wonder how they might survive in a radiation environment. In addition, as the cold war intensified in the early 1950s during the Korean War, the military became increasingly concerned about the possibility of fighting a tactical nuclear war and how its soldiers would operate on an irradiated battlefield.[15] These concerns led to greater government support for studies on the effects of radiation on military (and civilian) populations. Second, the culture of postwar clinical trials provided the space in which various kinds of experiments with human subjects were not only feasible but also permissible. As we will see in chapter 3, in the immediate postwar period, the Atomic Energy Commission rejected early attempts to use healthy volunteers for military radiation experiments, and researchers turned, instead, to cancer patients, particularly those with advanced and disseminated diseases. By the mid-1950s, physicians had considerable access to patients for these studies since, with the growth of clinical trials, cancer patients with advanced disease were commonly the first site on which to test almost every kind of speculative, and often toxic, therapy. Third, and perhaps most important, following the atomic bomb attacks on Hiroshima and Nagasaki, radiation researchers had identified a new disease, the so-called radiation syndrome or radiation sickness. It became a paradigm problem that attracted researchers from diverse disciplines, especially those with interests in radiation biology, military medicine, and cancer therapy. Radiation sickness provided a common ground on which researchers from these various disciplines could study different aspects of the radiation syndrome and readily exchange ideas and techniques. Along with these developments, physicians had, for some time, been using total- and partial-body irradiation to treat patients with disseminated disease, particularly lymphoma and leukemia. Since, as I already mentioned, cancer patients with advanced disease had became an important site for carrying out speculative research, it was not a large step for physicians to treat patients with disseminated cancers with total- or partial-body irradiation while investigating questions of interest to the military. Consequently, cancer patients became the preferred site for investigators to work on the paradigm problem of radiation sickness. At one and the same time, radiation sickness was a common toxicity in TBI treatments for cancer and a necessary syndrome for answering military questions.

Although a number of investigators participated in these military experiments, the Saenger case has received the most attention. One of its most striking features is that, in spite of its notoriety, it has remained the subject of intense controversy for more than thirty years without any closure at any point. With rare exceptions, ambiguity and uncertainty have marked the various investigations into Saenger's program. There are a number of reasons why this case has proved so intractable, and I will address them in the course of this book. One of the most significant of these reasons was the difficulty Saenger's contemporary peers had in judging his research since his practices looked and felt too much like their own. His research shared so many of the attributes of medical investigations of the cold war period — both its normal and its egregious characteristics — that it was difficult for his peers and for later critics to draw a sharp boundary between his practices and those of others. Consequently, they could not precisely identify those characteristics that would mark his program as clearly unethical. This difficulty suggests that a close scrutiny of Saenger's program not only provides insights into the research practices of the postwar period but actually lays bare many of the conflicts and tensions of clinical research that may not be as apparent in more conventional medical research of the period.

For example, the Saenger case highlights the inherent tensions between research imperatives and therapeutic necessities since Saenger's research goals were, in large part, aimed at answering military questions while his clinical program addressed radiation treatments for advanced cancers. Because the research and therapy issues clearly addressed different medical spheres, the conflict between them comes out in sharp contrast. Yet the tension between research and therapy is not particular to Saenger's program; rather, it is part and parcel of all clinical studies. I vividly remember my first exposure to this inherent conflict in the early 1980s when I began a postdoctoral physics traineeship in radiation medicine at Thomas Jefferson University in Philadelphia. A planning conference was held every morning (at an ungodly early hour): each new patient was presented by one of the residents, and the presentation was followed by a discussion of possible treatment plans by the attending physicians. These conferences were mundane and tedious except on the rare occasion when the chairman, Simon Kramer, was present. Kramer was one of the leading figures in radiation therapy in the postwar period and a great proponent of scientific medicine supported by randomized clinical trials. New patient conferences with Kramer were always a treat since the level of discussion was quite high and leavened—as always—with his sparkling wit. I clearly recall a conference at which one of the attending physicians declared that he did not intend to enroll the patient under discussion that morning in an available clinical trial since he was uncomfortable with one of the arms of the study. Kramer patiently asked about what evidence the physician had to support his position. The physician could not provide any published evidence. Kramer sighed and remarked something along the following lines: "My dear Doby,[16] I love you, but you are not making any sense—all you are telling me is that you don't like one of the arms of the study. I am hearing personal sensibilities and not hard evidence. If you don't have such evidence, you have a duty to enroll the patient in the trial."

I remember a second incident about ten years later. By then, I was the chief of the Clinical Physics Service at Memorial Sloan-Kettering in New York, and I was discussing the possibility of Memorial participating in a multicenter clinical trial with the physician-in-chief, Samuel Hellman. I was taken aback by Hellman's outright dismissal of the idea. This was the same individual who had, while chair of the Joint Center for Radiation Oncology at Harvard, overturned the then-prevailing American orthodoxy of using radical mastectomy for treating primary breast cancer. He had shown that radiation therapy following local excision of the tumor had comparable survival rates, and often had better cosmetic outcomes, than radical mastectomy. Of all people, I thought, how could he not support clinical trials? Hellman was emphatic. He ran through many arguments against randomized trials, including his concern that participating physicians are not given access to the accumulating results of the study. They must continue to enroll patients while remaining blind to the growing evidence. Hellman was very uncomfortable with this position since it was, he believed, the duty of each physician to make a judgment for every patient that was based on all the available evidence, including anecdotal data as well as the results accumulating in ongoing clinical studies. For Hellman, the physician's fiduciary responsibility to the patient trumped research imperatives.

These two anecdotes encapsulate the inherent tensions at the heart of clinical studies. Such tensions are by no means simply a product of postwar concerns. The conflict between advancing science and protecting research subjects was essentially no different for Pasteur in the mid-nineteenth century than it was, for example, for the medical researcher Donnal Thomas during the cold war, as we shall see in chapter 3. What changed was that ethical concerns had become (certainly by the time of my training) a quotidian issue that was debated each morning at planning conferences.[17] For researchers like Simon Kramer, the path through this ethical thicket was guided by randomized trials; for Samuel Hellman, his responsibility to his patients overrode the research imperatives for which randomized trials are designed. There is no universal solution to this conflict. Nevertheless, in their day-to-day activities, investigators often lose sight of this underlying conflict as they seek to enroll patients in their studies. Since most clinical trials are about controlling the disease from which the patient is suffering, it may appear to investigators that the trial necessarily satisfies both the research aims of the study and the therapeutic needs of the patient. In the Saenger case, it is not possible to ignore or paper over these tensions; the military goal of the research (studying radiation sickness in soldiers) is in an entirely different category from treating advanced cancers.

The Saenger case also reveals another troubling aspect of clinical trials, the use of research subjects as proxies. In a paper written almost forty years ago (and still perhaps the most penetrating discussion of clinical ethics), the philosopher Hans Jonas discusses a number of issues in human experimentation. He devotes one section to what he calls the *peculiarity* of human experimentation. Jonas argues that, unlike in other forms of experimentation, in clinical studies "[we] must operate on the original itself, the real thing in the fullest sense, and perhaps affect it irreversibly." But the most disturbing characteristic of clinical research is, for him, that patients become mere samples: "What is wrong with making a person an experimental subject is not so much that we make him a means . . . as that we make him a thing . . . a passive being merely to be acted on. . . . His being is reduced to that of a mere token or 'sample.'"[18] Even soldiers drafted to fight in a war, Jonas argues, are in a better position, for, unlike patients, they still retain their agency. Each patient in a clinical trial becomes a thing that stands in for something else, namely, a member of a cohort used for answering the research questions addressed by the trial. But the troubling issue of patients being reduced to mere samples is usually lost sight of by investigators who are under pressure to enroll and treat them. In the Saenger case, the use of patients to stand in for soldiers on the nuclear battlefield is laid bare. It reminds us that the use of proxies is, in its broadest sense, not peculiar to Saenger's studies but a common feature in human experiments.

One of the goals of this study is to characterize clinical research during the first half of the post–World War II period. In contrast to most other histories of human experimentation, which articulate idealized accounts of medical research (accounts to which I return below), this one emphasizes material practices as they operate in the clinic. This approach relies on investigations from the sociology of scientific knowledge or science studies.[19] It also draws on my more than twenty years of experience in research and in the clinic in radiation medicine. In this study, scientific research is characterized as deeply embedded in social, political, and cultural networks, the production of scientific knowledge as uncertain and subject to contingencies, and, consequently, knowledge claims as often highly contested. It is only after consensus is reached that scientists retrospectively write histories (in scientific papers and grant applications) that banish controversies and replace them with a story plotting the inevitable emergence of the newly reified knowledge. The neat, well-ordered, and teleological character of scientific research so typical of many accounts is a retrospective construction. In a similar way, clinical conduct arises from the interplay of similar social, political, and juridical forces. The resulting system of ethical practices has a history: it is the product of contingent events, and it has a dynamic and local, rather than universal, character.[20]

Since this study emphasizes material practices over idealized accounts of medical experimentation, it differs markedly from most histories of human experimentation. A composite picture of these accounts (somewhat caricaturized, I admit) would contain the following elements. To begin with, they are almost always what we might term *essentialist* in character.[21] Research has an essence that can be characterized prior to understanding how it is practiced in the laboratory and the clinic. It is not a messy affair filled with controversies and contingencies but, rather, a well-ordered enterprise, one that thrives in an environment freed of social and political considerations. And, even when social and political forces intrude, they only guide the directions of research; the knowledge claims that are produced remain independent of those influences.[22] All too often, essentialist histories do not even explicitly characterize the essence of medical research but, rather, assume that it is simply the opposite of what they identify as unethical behavior and poor practices. Likewise, many essentialist accounts are predominantly influenced by contemporary bioethics, which assumes that histories of human experimentation should be viewed through the lens of its *universal* ethical principles.[23] Furthermore, those ethical principles arise primarily, not from the interplay of political forces, but, rather, from an understanding of what is common to ethical behavior in all people.[24] Thus, ethics stands outside medical research and outside all politics so that all judgments (including historical ones) can be rendered through the rational application of universal principles.

In essentialist accounts of human experimentation, research is presented in this highly idealized fashion and contrasted to the research efforts of "suspect" investigators. Invariably, their studies are judged as falling far short of the idealized standards, and their research declared pathological. For example, if research directions change throughout the course of their studies, or if research goals originally envisioned are not realized, then the investigators may be accused of bad science and, consequently, unethical behavior. Or, if their consent statements are not ideal, or if they did not use consent at all, the suspect investigators may be condemned for unethical behavior. Indeed, as we will see in the last chapter, ACHRE considered researchers morally blind if, prior to government regulations on informed consent, they did not follow what have become current consent practices.

My use of a suspect case to lay bare some of the disturbing aspects of clinical investigations would be anathema to such idealized histories. Indeed, the tenor of essentialist accounts is one of exposé and reform. Suspect research like Saenger's is revealed as pathological and segregated from normal investigations. Suspect studies tell us nothing about normal or "good" science—in fact, they are worthwhile only for identifying egregious or "bad" science. In addition, essentialist accounts often call for the introduction of further ethical rules to control medical research. Rather than fully coming to terms with the fundamentally problematic nature of clinical studies, these essentialist histories envision a very different world. Human experimentation is an unquestioned good that advances human welfare. Unfortunately, unethical studies sometimes arise, but—as the argument goes—they are few in number and can be controlled through the introduction of new and improved ethical rules. In the end, human experimentation is domesticated. We can have advances in medicine without human cost.

These opposing views of scientific research and ethical conduct lead to divergent interpretations of the archival record. Essentialists might criticize Saenger because medical orthodoxy now believes that he could not effectively treat certain types of tumors with the doses of TBI he delivered, or because he failed to produce a biological dosimeter, or because he mixed so many goals. By contrast, from the perspective presented here, failed efforts, changing directions, and mixed goals might be nothing more than typical science in the making. Essentialists would also explain the clinical conduct of practitioners according to universal bioethical standards, while I would consider the regulation of research practices as the product of agreed forms of conduct within the broader medical/government community. For example, the consent requirement contained in government regulations introduced in the mid-1960s has been identified by some bioethicists as an early formulation of universal autonomy rights, while I find that consent was introduced to try to reduce the levels of risks to patients, investigators, and the NIH in order to maintain the medical research enterprise.[25]

This book explores the complex and mutually supportive relationship between clinical ethics and medical research practices during the cold war as well as the symbiotic relationship between military medicine and cancer therapy using TBI. Although I concentrate on the research program of Eugene Saenger, in order to place his work in perspective, I also look at the work of other postwar researchers: Bernard Fisher, for the epistemology and practice of cancer clinical trials; James Shannon, for changing the ethical governance of clinical investigations; and Donnal Thomas, for the complex relationship between radiation therapy and military medicine.

Such case studies provide a detailed perspective on the research practices of postwar investigators while also revealing important clues to their clinical conduct.[26] It is through such a forensic investigation of the programs of some of the cold war investigators that the case method may reveal important facets of research practices that are not available for scrutiny in historical accounts that rely on idealized versions of medical research or refer to community standards of practice. I have already mentioned that critics of Saenger failed to effectively judge his ethical behavior when they attempted to compare his research to the standards of his peers. The case method provides us with another method of assessment. Along these lines, Carlo Ginzburg reminds us that the famous Italian art critic Giovanni Morelli was able to assess the attribution of a painting, not by considering the painter in relation to his circle (i.e., community standards), but, rather, by focusing on "the most trivial details that would have been influenced least by the mannerisms of the artist's school: earlobes, fingernails, shapes of fingers and toes."[27] Case studies, then, are capable of providing us with evidence that produces knowledge that is "indirect, presumptive, conjectural."[28] In using the case method for studying Saenger and his coinvestigators, minute details and marginal data—revealed in the investigators' particular use of language, their behavior with colleagues and others, and the style and content of their publications—provide important clues to their character and actions.

From the accumulated evidence presented here in the case studies, readers will, I expect, reach their own judgments regarding the ethical character of the researchers. And, although I comment in places on their behavior, especially Saenger's, it is not my purpose to attempt an ethical assessment. Readers who are looking for a condemnation of Saenger, or, for that matter, for his rehabilitation, will have to go elsewhere. Indeed, as I have already discussed, there is no shortage of ethical judgments. Some of the critics have assigned political and personal motives to Saenger alone while reserving the mantle of truth for his interlocutors. Others have attempted to answer one or a combination of questions, among them: To what extent was TBI appropriate treatment for disseminated cancers at the time? Was the level of complications acceptable? Were the consent statements sufficiently revealing? Were the rules governing completion of the study appropriate? Can current ethical rules be used to judge earlier research? and so on. Still others have tried to compare Saenger's research practices to those of his peers. If I were to participate in this plethora of ethical assessments, I would become a participant in, rather than an observer of, the history of these events and cloud the story. Rather, I have chosen, following methodologies from the social studies of science, to provide a balanced (i.e., symmetrical) account of the Saenger controversy.[29] I believe that, by adopting this approach, I recount a story that provides a more revealing (and disturbing) version of the Saenger case.

In addition, if I were to attempt to produce a robust ethical assessment, I would find myself undermining the very argument that I have been making, namely, that the inability of Saenger's interlocutors to reach a definitive judgment says something fundamental about the character of medical research itself (e.g., that it is highly contingent and dynamic). By my count, there have been at least nine separate investigations of Saenger's program, beginning with his first peer review in 1966. All have led to ambiguous and unsatisfying judgments. These difficulties, however, should not be interpreted as simply a consequence of the limitations of his interlocutors; rather, we should recognize that Saenger was able to combine the egregious and the normal components of the spectrum of research practices so that they could not readily be separated. As I have already argued, it is this characteristic of the Saenger case that provides us with a path into the thicket of postwar clinical trial practices.

The book is divided into three parts. In the first part, I provide a framework for postwar medical research practices, focusing on clinical trials, ethical regulation, and TBI studies. In chapter 1, I explore the mid-1970s work of a distinguished oncology researcher, Bernard Fisher, in order to bring out a number of salient characteristics of clinical trial culture. In particular, Fisher's multicenter trials were governed by a tightly regimented system of practices that were capable of producing highly reliable results. At the same time, utilitarianism and the principle of equipoise provided the ethical framework to support the trials. The former principle regarded weighing (and often favoring) the benefits of research against the risks to patients. The latter principle was particularly important for entering patients in a clinical trial since clinicians believed that they were behaving ethically if they were equally ignorant of the possible outcomes of the standard and experimental treatments on offer. In spite of the highly ordered nature of the multicenter trials that were capable of producing statistically significant results, Fisher and other investigators encountered great difficulty in having their knowledge claims accepted by their research peers. Consequently, by the 1970s, the medical research landscape was filled with a number of new entities—consensus conferences, ad hoc committees, and innovative statistical techniques—all intended to bring order to the many controversies that often accompanied clinical trials. Of course, not all clinical trials were controversial, but the machinery for settling controversy that was put in place attests to the importance that this issue attained. In spite of these collaborative and institutional programs, clinical conduct during this period continued to be governed primarily by the individual judgments of physician-investigators.

In chapter 2, I show that, by the late 1950s, equipoise—along with a utilitarian calculus—was no longer sufficient to govern experiments with human subjects. In the early postwar period, clinical research had moved toward large-scale, highly engineered studies, while the dramatic growth in government support led to an increasing number of investigators and a wider range of human studies. At the same time, physician-investigators were increasingly exposed to legal suits with (for all intents and purposes) neither case law nor formal codes of practice to protect them. By the late 1950s and early 1960s, a series of highly publicized press reports, epitomized by the thalidomide scandal—where there were revelations that women who had used thalidomide to control morning sickness were giving birth to deformed children—highlighted the difficulties the medical community had in controlling human research. In this chapter, I focus on the work at the NIH of James Shannon, who, by the late 1950s, had begun to fear that the medical research enterprise—and the NIH itself—was in danger owing to the unethical behavior of some physician-investigators. He and others at the NIH had come to believe that, if researchers were subject to a review by their peers, the risk to investigators, patients, and the NIH would be reduced. By 1966, the NIH put in place a system of regulation that required all proposals for government support to receive the prior approval of a local peer review committee charged with assessing the risks and benefits of the research as well as the appropriateness of consent. The system that Shannon put in place was a radical change in the governance of medical research. Trust would now reside, not in the prerogatives of physician-investigators, but in the judgment of more transparent medical/government institutions.

In chapter 3, the last chapter of the first part, I consider the intimate relationship between research for cancer therapy and military medicine. I follow the efforts of Donnal Thomas from 1957 to 1977 and his development of a successful therapy for leukemia with TBI and bone marrow transplantation. I consider how his work drew on and also contributed to methods for treating the victims of nuclear explosions who were suffering from radiation sickness. Thomas emerges as but one of a number of investigators who operated across very diffuse and dynamic boundaries between military medicine and cancer therapy. His research also exemplifies the peculiarity of clinical investigations. Early on, his research goals were anything but clearly defined as he wavered between treating radiation sickness after nuclear attack and controlling childhood leukemia. His treatments for leukemia were highly toxic since he inevitably tested every new and potentially lethal strategy on his patients, yet his publications do not indicate that he used formal consent practices until well over a decade after he initiated human studies. At the same time, Thomas appears to have been deeply affected by the struggles of his patients, and his efforts did eventually lead to long-term survival rates in childhood leukemia and even a Nobel Prize in medicine. Perhaps it was his ultimate success that has shielded him from essentialist assaults on his ethical practices.[30]

The second part consists of four chapters specifically focused on events in Cincinnati between the late 1950s, when Saenger launched his project, and the early 1970s, when his program was exposed to intense public scrutiny and, ultimately, closed down (in 1972). Chapter 4 focuses primarily on the early stages of Saenger's program, from 1959, when he first submitted a proposal to the military to study radiation effects, until ca. 1965, by which time he had initiated and greatly expanded his research program. The chapter identifies a crucial event, namely, his decision to substitute TBI for chemotherapy in treating cancer, that provided Saenger with the patients he needed as proxies for soldiers. In addition to his search for a human dosimeter to register the amount of radiation a soldier had received, he also investigated the psychological effects of TBI in order to understand how effectively a military commander could operate in a nuclear environment. The chapter shows that research programs often have to be modified as ideas and techniques develop to meet changing circumstances. For example, Saenger and his coinvestigators found that they could not deliver the high doses of radiation they sought because of the toxicity. As a consequence, the program changed in fundamental ways: they began to launch an effort to use bone marrow transplants to mitigate the radiation effects, and they also had to look for other indicators of radiation damage because they were modifying the patients (i.e., the dosimeter) with a bone marrow transplant. Such changes in direction do not necessarily reflect bad science; rather, they are rather typical of research as it develops. At the same time, the troubling aspects of the program also emerge, particularly Saenger's continued treatment of his patients with TBI before he could effectively use a bone marrow transplant to mitigate the toxicity as well as the callousness of his psychological researchers, who performed psychomotor tests on debilitated and dying patients.

Chapter 6 investigates the role of ethical peer review at Cincinnati and shows that contemporary judgments of Saenger were highly contingent and influenced by social and political forces. The chapter follows the changing strategies that the local peer review committee adopted to assess Saenger's research proposals. In an early submission, the committee was deeply suspicious of the research program, yet it felt uncomfortable judging its risks and benefits. The committee members understood that, if they were to assess benefits, they would have to evaluate the scientific merits of the research, judgments that were, they believed, outside their areas of expertise. The committee, however, was able to (provisionally) approve the program, but only after it received the personal assurances of a cancer specialist. In a later review, although the committee was again deeply concerned about the aims and toxicity of Saenger's research, it had grown more comfortable with judging proposals. It believed that it could separate out the ethical issues from the scientific aspects by focusing on the content of the consent statement. The committee was, therefore, able to approve the program on the basis, not of personal assurances, but of the narrowest of technical grounds. The chapter also shows that the concerns of the peer review committee, combined with the changing interests of Saenger's coinvestigators, moved the program in a direction that embraced more overtly clinical goals. Rather than putting a halt to Saenger's program, the committee, in its institutional role, sought to modify the research and tame its more disturbing characteristics.

Chapter 7 examines a number of issues of public disclosure, the role of investigative government, and the reactions of institutions and individuals to investigative government's queries. It follows the response in 1971 of the University of Cincinnati administration to sensational reports in the press regarding unethical experimentation that presented Saenger as a purveyor of mad science. Initially, the Cincinnati team attempted to control the furor by producing an alternate story portraying Saenger's research practices in a highly idealized fashion. Continuing public pressure, however, forced the administration to seek reviews of his program by internal and external committees. All were mildly critical but generally supportive, although a noncommissioned report by some of the junior faculty at the University of Cincinnati, which was ignored by the administration, accused Saenger of killing a number of patients. In addition, the medical center was forced into a bitter struggle with Senator Edward Kennedy over access to the patients. The Cincinnati administration prevailed by turning to the powerful argument of protecting the patients' welfare and confidentiality. Saenger's program was finally closed down, not because of concerns about the ethics of his experiments per se, but because the administration was further embarrassed by revelations that Saenger had secretly negotiated a contract renewal with the military in the midst of the investigations. The chapter highlights two related arguments: that the fluidity of Saenger's program led different investigative committees to read his research in different ways and that medical prerogatives play an important role in shielding the medical community from outside pressures.

Whereas chapters 4, 6, and 7 are presented through the perspective of Saenger, his peers, and his critics, chapter 5 offers an account of one of the patients, Maude Jacobs. Here, I draw on my years of experience as a radiation oncology physicist to imaginatively reconstruct from sparse archival sources a "firsthand" account of her experiences. We follow her through a frightening radiation treatment, the appalling complications that ensued, and her stoic efforts after returning home to provide for her family in the few weeks of life that remained. Without such a story, we have no way of understanding what Sanger actually did to patients; most of the available accounts regarding the effects of his program are buried in statistical tables of survival and toxicity—data that have been used effectively by both Sanger's supporters and his detractors.[31] Moreover, since Maude Jacobs's story is narrated from her own perspective, she emerges, not simply as a patient ground up in the medical machine, but also as an individual who shows strength and character in the face of great suffering.

In chapter 8, which constitutes the book's final part, I return to the question of why there has been so much ambiguity in judging Saenger's work. In the mid-1990s, Saenger was investigated by a number of government bodies, including ACHRE. The political and cultural climate in which he was judged differed substantially from that of the early 1970s, when various internal and external committees had investigated his research program. One aspect of this change was that, by the 1990s, bioethics had clearly established itself as the most authoritative voice of ethical conduct: the choice of a bioethicist to head an advisory committee investigating cold war research is a mark of the prestige and power that bioethics had attained. After briefly tracing the development of bioethics from the late 1970s through the mid-1990s and the role of neoliberalism in shaping it, this chapter follows in detail ACHRE's deliberations in the Saenger case. It notes that the committee was badly split following attempts by its ethicists to apply universal bioethical principles to make retrospective judgments of Saenger and other researchers. However, when the committee attempted to compare Saenger's research to that of his contemporaries, it could not distinguish his practices from his peers' and could reach only an ambivalent judgment.

In a brief epilogue, I follow the Saenger case to a civil suit brought by some of the families of Saenger's patients, which again did not lead to closure (although it was settled out of court in 1999). The families received minimal monetary compensation, and, while the university agreed to place a memorial plaque on its campus, Sanger did not show any contrition or apologize to the families. I conclude by returning to the sentiments of the third epigraph to this introduction, namely, that physicians and scientists in modern medicine have entered a "no-man's-land" that was once the sole purview of the sovereign. The epigraph can be taken to mean that physician-scientists have assumed authority over the lives and well-being of patients. More important, it can also be taken to mean that the prescriptive rules governing ethical research are general enough that they have opened up a no-man's-land in which the egregious and the normal can operate—even, at times, in the same clinical experiment. Research practices like Saenger's, which were highly fluid affairs, could operate in the interstices between the generalized ethical rules. In turn, ethical regulators tried to fix such investigational enterprises, to hold them in place so that they could categorize and judge them according to their prescriptive regulations. The battle between human experimentation and ethical regulation was one between fluid research and static codes. But Saenger's program was anything but a fixed entity. It was a dynamic social enterprise that effectively evaded the net of his critics for decades.

Saenger's research program operated in a clinical trial community that recognized it as one of their own. To understand the cultural milieu of that community, I turn in chapter 1 to a discussion of the epistemology and practice of cancer clinical trials. In chapters 2 and 3, I consider the governance of clinical conduct and the intimate relationship between cancer therapy and military radiation research.

Notes

1. Rosenberg, *Care of Strangers*, 8–11, 142–65; Starr, *Transformation of American Medicine*, 154–79; Stevens, *In Sickness and in Wealth*, 159–70.

2. Marks, *Progress of Experiment*, 17–97.

3. Murtaugh, "Biomedical Sciences."

4. Patterson, *Dread Disease*, 248–51.

5. Bud, "Strategy in American Cancer Research," 426–31.

6. Leslie, *Cold War and American Science*, 172–74; University of Texas, *First Twenty Years*, 212–23.

7. For the mobilization of science and the state, see, e.g., Hughes, *American Genesis*.

8. Laszlo, *Cure of Childhood Leukemia*.

9. Abeloff et al., "Breast."

10. Bailar and Smith, "Progress against Cancer"; Patterson, *Dread Disease*, 251–53.

11. Risse, *Mending Bodies*, 155–86.

12. See chapter 2 and the references there.

13. Faden and Beauchamp, *Informed Consent*, 4–9.

14. ACHRE, *Final Report*, 236–39.

15. For further details and references, see chapters 3 and 4.

16. Kramer was using a diminutive for Ralph Dobelbower, one of the young attending physicians at Jefferson, who went on later to chair his own department.

17. For a fascinating look at Pasteur's ethical dilemma and his actions, see Geison, *Private Science*.

18. Jonas, "Philosophical Reflections," 2, 3.

19. I have relied primarily on Callon, "Sociology of Translation"; Collins, *Changing Order*; Latour, *Science in Action*, and *Pasteurization of France*; Pickering, *Mangle of Practice*; and Shapin and Schaffer, *Leviathan and the Airpump*.

20. I have been influenced by the work of Alasdair MacIntyre, especially *After Virtue*, on the importance of understanding the historical context of ethical practices.

21. For essentialist interpretations of the Saenger studies, see, e.g., Little, "Experimentation with Human Subjects"; Stephens, "Medical Research on Humans"; Lawrence Elish, "Legal Rights of Human Subjects in the University of Cincinnati Whole Body Radiation Study," January 11, 1973, DOD 042994-A;8/16; David Eligman, "Statement to House Judiciary Committee, Subcommittee on Administrative Law and

Government Relations," April 11, 1994, DOD 042994-A;13/16; ACHRE, *Final Report*, chap. 8; and, most recently, Stephens, *Treatment*. For typical essentialist accounts of other examples in human experimentation, see, e.g., Hornblum, *Acres of Skin*; Moreno, *Undue Risk*; and Welsome, *Plutonium Files*. An exception is Goodman et al.'s *Useful Bodies*, which attempts to place human experimentation in the context of the state. Although many of the essays in this collection still follow an essentialist position, some of them take a more material and pragmatic approach. For a more extended discussion, see Kutcher, review of *Useful Bodies*.

22. Merton, "Normative Structure of Science." For the critical influence of politics on the directions of scientific research, see Dickson, *New Politics of Science*; and Greenberg, *Politics of Pure Science*.

23. For the basic principles, see Beauchamp and Childress, *Principles of Biomedical Ethics* (1994), 37.

24. ACHRE, *Final Report*, 116–17.

25. For more details, see chapter 2.

26. For a discussion of the uses of case studies, see Forrester, "If p, Then What?" For the case method in ethics, see Jonsen and Toulmin, *Abuse of Casuistry*.

27. Ginzburg, "Clues," 97. Ginzburg also points out that Sherlock Holmes (Conan Doyle was a contemporary of Morelli's) relied on minute evidence, imperceptible to most people, for clues to reveal the perpetrator of the crime. And Freud later appropriated Morelli's idea that small, inadvertent gestures reveal our character far more than formal gestures do (ibid., 97–99).

28. Ibid., 106.

29. Bloor, *Knowledge and Social Imagery*, 175–79.

30. Allan Brandt points out that, for some critics, research is ethical if successful and unethical if a failure ("Polio, Politics, Publicity," 452).

31. Stephens's *Treatment* is an exception. Stephens has written movingly about the Cincinnati patients, although her discussion of Maude Jacobs differs from mine.

PART I

RESEARCH IMPERATIVES AND CLINICAL ETHICS

1 Cancer Clinical Trials

Termites build their obscure galleries with a mixture of mud and their own drop-
pings; scientists build their enlightened networks by giving the outside the same
paper form as that of their instruments inside. In both cases the result is the same:
they can travel very far without leaving home. • Bruno Latour, *Science in Action*

....

In the post–World War II period, the randomized trial became the gold
standard for answering questions in cancer medicine in the United States.
By 1984, Bernard Fisher, one of the leading cancer researchers of the postwar period, could readily claim that prospective randomized trials were
one of the most significant advances in medicine: "They apply the scientific method to clinical problem solving, and provide the mechanism for
obtaining, with as little bias as possible, answers to important clinical and
biological questions that could not be obtained in any other way."[1] As
early as the 1950s, American cancer researchers were struck by the ability
of clinical trials to answer medical questions, to resolve controversy, and
to advance medicine. These investigators believed that, with the apparatus
of a clinical trial, impartial observers using objective criteria would establish medical facts that would replace medical knowledge passed down by
tradition. Doctors would, thereby, gain more authority in the eyes of their
patients. Antiquated attitudes would be swept away and replaced by a
rational medicine based on unbiased prospective randomized trials.

How did the randomized clinical trial come to dominate postwar medical investigations? And what were the standards of practice that supported
these trials? The first section of this chapter follows the growth of cancer
clinical trials from the end of the Second World War until the early 1970s
and President Nixon's War on Cancer. To understand the development of
clinical trials, it is necessary to follow the chemotherapists who were the
driving force behind the introduction of the clinical trial apparatus early

in the postwar period. By the time of the War on Cancer, they dominated the National Cancer Institute (NCI), and the features of cancer clinical trials had been established and stabilized.[2] From the outset of the postwar period, American clinical trials found an exemplar in the work of Bradford Hill and the British Medical Research Council (MRC).[3] Most of the elements that were taken over by the Americans had been initially developed during the British studies: random allocation of patients to competing treatment arms, formal protocols, measurable end points of clinical efficacy, the null hypothesis, and statistical measures to establish whether the treatment arms differed.

These standards of practice are exemplified in Fisher's mid-1970s study (discussed below) of the use of L-phenylalanine mustard (L-PAM) and other chemotherapy agents in the management of primary breast cancer. Fisher's trials exemplify how randomized multicenter trials were designed and implemented as well as their strengths and weaknesses. The strength of clinical trials depended on their size and randomized structure, which provided the capability of producing unbiased and statistically significant differences between the various treatment options on offer and, thus, answering important clinical questions. Their weaknesses arose from problems in enlisting cohorts of physicians and patients, the political and economic forces limiting the choice and scope of the various arms of the trials, and the fiercely contested nature of their knowledge claims. But, to Fisher and others, clinical trials provided more than a powerful utilitarian technology allowing physicians to choose the best care. They were also viewed as a laboratory tool to study cancer biology. Animal experiments would provide important clues, which could then be tested and elaborated in human trials. As Fisher put it: "It is hypothesis, and not individual experiments, that require testing in the human."[4] Clinical trials were part of a research culture that extended from clinicians, who participated in the program and enlisted patients; to clinical researchers, who appropriated funds and designed and managed the trials; to laboratory researchers, who sought new drugs to test in the clinic.

In spite of the huge apparatus and resources that were brought to bear on clinical trial programs, the knowledge claims of the studies were often controversial. The third section of this chapter focuses on the difficulty researchers had in reaching consensus. The gaze of the cancer researchers was so focused on the unbiased and objective design of clinical trials that they underestimated the degree to which clinical trials could be challenged on any number of grounds.[5] These attacks came from multiple directions, including researchers, prominent physicians, clinicians (physicians predominantly involved in clinical duties), and special interest groups. To counter the endless debates, new methodologies like meta-analysis were developed and new social structures like consensus conferences put in place by the National Institutes of Health (NIH) to try to settle controversy and construct medical knowledge. Yet, despite the promise of their highly formalized and supposedly unbiased character, the NIH consensus conferences often had difficulty reaching closure, and, when they did, the basis of agreement might depend as much on social and political forces as on rational arguments and scientific evidence. Thus, clinical trials could not readily produce medical knowledge through a purely rational process, as idealized pictures of medical research would have us believe. To be sure, not all clinical trials were controversial. Some were accepted without dissent, while others were simply ignored. In a few studies, the results were so convincing that the new procedure was rapidly accepted by the medical community, even in some cases without the benefit of a randomized trial (e.g., Donnal Thomas's bone marrow transplant study [discussed in chapter 3]). Nevertheless, when the stakes were high and the statistical distance between the arms of the trial narrow (as in so many cancer clinical trials), the results were often fiercely contested. Why else was so much apparatus put in place to try to settle clinical trial controversies? Many researchers had to travel over a long road littered with contingencies as they attempted to bring order to a discordant and highly variable medical community.

The Rise of the Chemotherapists and the Randomized Clinical Trial

The young researchers who congregated at Johns Hopkins University after the war, having been schooled in the wartime clinical trials environment, were more than receptive to the randomized trials of the British MRC. One of those young investigators, Gordon Zubrod, recalled their early enthusiasm: "At this time these studies of streptomycin in pulmonary tuberculosis by the Medical Research Council were published and had a profound influence on the Johns Hopkins Group."[6] Zubrod and others applied the randomized approach to study the value of penicillin for infectious diseases. At the same time, another young researcher, Louis Lasagna, turned his efforts to the active metabolites and struggled over clinical trial methodology, including subjective bias and the role of blind studies. When Zubrod went to the NCI in 1954, he carried with him a copy of Lasagna's report on clinical trials.[7] As soon as he broached the subject of randomized trials, the NCI took up his suggestions, almost on the spot.

If Zubrod had gone to the NCI earlier in the postwar period, he would have encountered a different reception. Until the late 1940s, the old medical guard who controlled the National Advisory Cancer Council, a body that guided the NCI, strongly resisted the type of large-scale engineered research that had become popular during the war.[8] But, in 1953, prominent physicians and others, among them Sidney Farber of Harvard, Cornelius Rhodes of Sloan-Kettering, and the philanthropist Mary Lasker,[9] successfully lobbied Congress for a $1 million grant to the NCI to investigate the possibility of an engineered program to cure leukemia. What evolved rapidly from this beginning was a three-pronged cancer program of drug screening, laboratory studies, and clinical trials. By the time Zubrod had arrived at the NCI, Rhodes had, with the support of the pharmaceutical industry, already screened something like twenty thousand drugs at Sloan-Kettering, but he could no longer keep up with the demands of researchers and industry. Rhodes's program was absorbed into the drug-screening section of the newly formed Cancer Chemotherapy National Service Center (CCNSC) and turned the NCI into a veritable pharmaceutical house. The massive volume of drug screening also required the NCI to initiate another large-scale program to produce tumor models and inbred strains of mice. The screening program likewise benefited from a large biostatistics group that was assigned to the CCNSC to develop experimental models for clinical and laboratory studies and to design a more efficient screening program.[10]

The CCNSC also administered a research program and a clinical studies section. It was the clinical panel of the latter section, with its leukemia mandate, that Zubrod found so receptive: "I called attention to the clinical trials in infectious diseases, such as tuberculosis and pneumococcal pneumonia, and to those involving analgesic drugs, and I suggested that these principles be tried in cancer."[11] It might at first seem surprising that Zubrod's presentation on the method of randomized trials did not lead to protracted internal battles but was, instead, readily received. His account should not, however, be treated as anything but an accurate description of events since his writings attest to his ability to dispassionately discuss the problems he faced and the pressures he was under at the NCI.[12]

Why, then, was Zubrod's suggestion taken up so readily at the NCI? In fact, why was there so much enthusiasm for clinical trials in the postwar period? Historians and sociologists have addressed this question in various publications. While their answers differ in details, they generally suggest that researchers took up clinical trials as a political strategy. Ilana Löwy, in a monograph on oncology research, puts the question in the following way: Why did the medical elite, legitimated by its expert status, surrender part of its power to support this new method of clinical judgment? Löwy argues that postwar structural changes in medicine, particularly the shift to public support, left the medical community exposed to public scrutiny and that the medical establishment turned to the clinical trial, with its objective measures, to ward off government pressures.[13] She also suggests, alternatively, that the development of clinical trials was a continuation of a struggle between biomedical researchers and medical practitioners for political control. Mark Sullivan, in a paper on placebo controls in medicine, makes the point that the medical community as a whole embraced the clinical trial because it offered legitimacy within the body politic through an "unbiased method open to all possible new truths and techniques."[14] Finally, Harry Marks emphasizes the importance of medical statisticians in transforming the conduct of clinical experiments.[15]

These analyses emphasize various threads of a fabric of epistemic, social, and political issues that were woven together so tightly that they could not easily be pulled apart. The war years, with their managed programs, produced researchers with an affinity for engineered medicine and clinical trials.[16] In addition, the NCI, with its skyrocketing drug program, was an environment that was geared to formally structured studies like the randomized trial. The chemotherapists also had an unmistakable enthusiasm for the value of drugs in cancer therapy, an enthusiasm that later took on an almost ecumenical fervor. At the same time, the chemotherapists were pressed into service by social forces, including congressional expectations and the demands of the pharmaceutical companies. As the chemotherapists gained strength and the scope of their studies expanded, their expectations for the role of chemotherapy grew. There is little doubt that they used clinical trial results, backed by the power of statistics, as a wedge to pry open the academic establishment and, perhaps, even to dominate it. The authority of the chemotherapists within the medical community and their legitimacy in the clinic were necessary if they were to gain control of patients and realize their aims of changing practices and curing cancer. Beyond gaining legitimacy within the medical establishment and power within the clinic, trials also provided the chemotherapists with a quantitative measure in their battles with Congress for additional funding. It would be difficult to separate the chemotherapists' use of quantitative measures to ward off government pressure and build up their medical enterprise from their genuine belief that more funds would lead to cancer cures.

Once clinical trials were accepted at the NCI as a means to answer cancer therapeutic questions, there still remained the crucial problem of recruiting a sufficient number of patients.[17] Since the NCI did not have adequate facilities of its own to carry out a full-scale leukemia trial, Zubrod collaborated with James Holland at Roswell Park in New York. This joint undertaking soon evolved into the Cancer and Leukemia Group B, the first of the many alliances of university medical centers.[18] The postwar multicenter clinical trial was essentially born with these cancer studies. The clinical panel, under the chairmanship of Zubrod, became the central point through which all protocols had to pass for review. It functioned as a training ground for multicenter studies as well as the central administration of the clinical research program.

The 1953 mandate from Congress to explore a leukemia program had, by the 1971 War on Cancer, extended to all cancers, as radiotherapy and surgery came under the NCI umbrella. Nevertheless, the chemotherapists' early strategic position at the NCI meant that they controlled substantial portions of its budget, even though they were still effectively outsiders in the medical establishment.[19] Vincent DeVita, later the head of the NCI, remarked on the earlier period: "Academic medicine, in general, rejected the intrusion of cancer chemotherapy and chemotherapists." Internists, who had run the first studies of leukemia, came into conflict with the hematologists, while surgeons managed almost all the other patients in the hospital.[20] Moreover, the chemotherapists believed that most cancers behaved as a systemic disease, a direct contradiction to the prevailing surgical view: this position placed them at odds with surgery, the most powerful of the medical disciplines at the time. In addition, unlike most surgeons, chemotherapists were more often found in the laboratory than in the clinic. Yet they had gained control of most of the funds, and they saw the clinical trial as a means to prove their claims and gain their place in the medical establishment.

Which areas of cancer therapy would provide the evidence needed to support the claims of chemotherapy as a treatment? The use of nitrogen mustard to treat leukemia was one possibility since it had initially demonstrated spectacular remissions, though, unfortunately, these were followed by wholesale recurrences. But, even if the chemotherapists cured all the leukemia and lymphoma cases, they would hardly budge the overall cancer death rate. They had to prove the efficacy of chemotherapy where the cancer incidence was high, namely, in solid tumors. Yet chemical cures, with one minor success (choreocarcinoma), seemed beyond the capability of their drugs. So the chemotherapists turned to what they termed *adjuvant* therapy; that is, they added chemotherapy to surgery or radiotherapy in order to control widespread (metastatic) diseases. This was a bold move, and almost all cancer therapy has since come under the purview of this new oncology medicine. The vast NCI budget was tapped to support the necessary laboratory research and clinical trials. Thus, by the 1970s, cancer clinical trials were overwhelmingly trials with chemotherapy (and they remain so). In a 1975 paper on adjuvant therapy in breast cancer, Fisher announced the proof the chemotherapists sought: "The most important aspect of the findings is . . . that for the first time it has been demonstrated from a well-controlled clinical trial that the rationale for using prolonged chemotherapy as an adjunct to operation is a sound one."[21] The principle that adding prolonged, and aggressive, chemotherapy to conventional local treatments with surgery and radiotherapy had improved cancer cures was proved. The future was not so much better surgery or radiation but an endless array of drug regimens that would lead to even better results. The future was for chemotherapy, and clinical trials were expected to show the way.

The Practice of the Postwar Multicenter Clinical Trial

This section examines a case study, Fisher's L-PAM trial and its follow-up studies, in order to understand the practice of clinical trials. Fisher began a 1977 paper with the hypothesis: "There is growing awareness that most if not all patients have disseminated disease at the time of diagnosis and that improvement in survival is only apt to result from employment of effective systemic therapy in conjunction with modalities used for local regional disease control."[22] This position was markedly at odds with what he labeled the *Halstedian hypothesis,* which had reigned for over seventy-five years, namely, that cancer was a local disease that spread steadily outward from its initial location and could be cured with local and regional surgery.[23] One of the proving grounds for this new and radical disseminated disease hypothesis, the fulcrum about which the future of adjuvant chemotherapy would turn, was the L-PAM study. This investigation is a classic example of a randomized clinical trial. It was framed with the standard clinical trial query: Is L-PAM with mastectomy better than mastectomy alone?[24] Everything in the design, execution, and analysis of the trial was grounded in the predominant ethos of objectivity and lack of bias. The patients were randomized only after they agreed to enter the study, thereby, presumably, rooting out any possibility of bias. In addition, the trial was double-blind[25] so that "neither the physician nor patient was aware of the treatment administered," thus "ensur[ing] lack of bias in subsequent observations."[26] If the physicians were not blind, they could inadvertently adjust their clinical evaluations on the basis of prior knowledge of the treatment given. And, if the patients were not blind, then "any change in a patient's symptoms" might be "the result of the therapeutic intent and not the specific physio-chemical nature of the medical procedure," that is, the result could be due to a placebo effect and not the activity of the drugs.[27]

In addition, physicians could enter patients into the trial only if they believed that there was no rational basis for choosing between the L-PAM and the placebo arms.[28] This principle of "equipoise" served at least two purposes. First, physicians would be less liable to unconsciously bias either patient selection or clinical evaluations if they did not favor one of the arms. Second, equipoise was a norm that physicians could apply to justify whether it was ethical for them to enlist patients into a trial. Indeed, unlike peer review and informed consent, equipoise was the one ethical regula-tion that arose directly from the practice of clinical trials. The statistical nature of a clinical trial was grounded in the null hypothesis that there was no difference between treatment arms. A statistically significant difference at the end of the trial would disprove the null hypothesis; that is, strictly speaking, a clinical trial proved, not that treatment A was better than treat-ment B, but that the null hypothesis (at some level of statistical confidence) did not hold. Equipoise was the ethical equivalent of the null hypothesis and helped grease the wheels of the clinical trial machine.[29]

Fisher chose survival and disease-free survival as measures of the effec-tiveness of L-PAM. Although the former measure was the sine qua non of cancer trials, it was not a practical end point since the possibility of tumor recurrence could extend out to ten years.[30] So Fisher used survival at two years without evidence of disease as a proxy, a decision that would later open his studies to criticism. In some important respects, the L-PAM study marked a departure from earlier cancer clinical trials. Fisher was trying to measure small survival differences because he was adding a drug on top of surgery. Consequently, in order to complete the study in a reasonable time, he recruited patients from a wide network of hospitals—some eighty-two institutions—belonging to various cooperative trial groups.[31] In addition, he designed the L-PAM study as the first in a series of interlocking trials in which the control arm of each was the test regimen in the previous trial so that the studies were successively calibrated to one another through their control arms. Not only were the drug regimens compared trial by trial, but the design also provided for a long-term drug escalation study in which patients would successively receive more toxic drug regimens until a maximum tolerable combination was reached, at which point the highest survival rates would, presumably, have been attained.[32]

To administer a program consisting of a large number of investigators participating in a series of interlocking trials, Fisher used a centralized and hierarchical structure. This design was a highly co-coordinated metrological enterprise with a statistical center controlling all the satellite institutions. The statistical (or co-coordinating) center was chaired by Fisher and staffed with statisticians, data managers, secretaries, and business administrators.[33] It had the infrastructure to administer the L-PAM study and to provide the continuity and the resources required for developing, funding, and pursuing further clinical trial studies.[34] This centralized body produced all clinical trial protocols and disseminated them outward to the various investigators at the participating hospitals. The co-coordinating center monitored the trial procedures to assure that the patients met the entrance requirements and that their diagnoses, treatments, and assessments were conducted according to the protocol. Even randomization was centralized. Once a patient agreed to enter the study, the participating physician was required to phone the co-coordinating center to determine the study arm, and, during this call, the "randomizer review[ed] a checklist of major eligibility requirements with the individual entering the patient to ensure that they have been met."[35] While all instructions flowed outward, all patient data moved back to the statistical center, where it was transferred from paper forms into a computer database system.

Although there was tight control, the trials operated effectively because of the built-in flexibility of the protocols. For example, the protocols could be subtly refined as they passed through the hands of a heterogeneous group of physicians, technicians, and nurses. At each step of the process, numerous undocumented but necessary and practical adjustments were made to meet local conditions. This subtly adjustable framework provided for medical practices that, construed broadly, were within written guidelines, yet it had sufficient flexibility to function at a number of diverse localities. At the same time, the overall integrity of the project was maintained by the co-coordinating center, which monitored a core set of study parameters. In addition, the participating physicians were able to ensure the viability of the program. The professional interests and political forces that led them to enter patients and run the studies according to protocol had to be quite robust since their clinical prerogatives and responsibilities were diminished in the multicenter culture, in which they were excluded from any knowledge of the progress of the trial.[36] Indeed, the statistical center was expressly organized to provide a "separation of patient care and evaluation functions" because of the possibility of "bias if study physicians are permitted access to the data during the course of the trial."[37]

The success of the program required Fisher to recruit physicians from successively wider domains. This was typical of most clinical research and, certainly, of all clinical trials. He enlisted a small group of collaborators who would participate in designing the study and contribute to the periodic analyses of trial results. This group would, in turn, enlist other investigators from a wider circle of university hospitals. At each of these hospitals, the investigators would recruit other attending physicians from a number of subdisciplines as well as numerous allied health personnel to carry out the study. The practitioners at the outermost domain would serve a critical role; they would enlist the patients, treat them, follow their progress, create the paper reports, and return them to the co-coordinating center.

The recruitment of patients was a problem that affected the L-PAM as well as other clinical trials. For example, it took over two years for the L-PAM study to accrue 370 patients, which clearly attests to the reluctance of the surgeons to enter patients into chemotherapy trials. In another trial of Fisher's, one that included mastectomy and lumpectomy arms, "physicians were reluctant to ask women to participate in a trial . . . which would result in a chance assignment to surgical therapies as diverse as 'having a breast removed' or 'merely having a lumpectomy.'" In this case, Fisher was forced to redesign the study and introduce so-called prerandomization, which permitted patients and physicians greater latitude than the statisticians would have preferred.[38] The great difficulty that clinical trial researchers had in recruiting patients led to a change in strategy by the late 1970s. Nonacademic community hospitals, which previously had not participated in clinical trials, were brought into the cooperative networks in order to recruit more patients.[39]

In his 1977 paper, Fisher claimed a victory for L-PAM and forecast that even better results lay ahead with more aggressive chemotherapy regimens. He based his conclusion on the finding that the disease-free survival rate at two years was 76.2 percent with L-PAM and 68.4 percent with placebo (at the $P = 0.009$ level).[40] Nevertheless, there was a disquieting note. The gains were predominantly a consequence of drug activity in a stratified sample of patients under fifty years of age, while it was the over-fifty cohort, with its higher rate of cancer, that was of greater importance.[41]

Fisher reported on the toxicity rather differently than he did survival. He either relied entirely on qualitative descriptions or, when statistics were used, complemented them with rhetorical descriptions. For example, he argued that complications were "limited" and "manageable" and that there was a high compliance of the patients to the drug regimen. Although 40 percent of the patients on L-PAM "experienced some degree of nausea and vomiting," Fisher characterized the complications as minimal, even though in 11 percent "the symptoms were greater."[42] This highly asymmetrical presentation between how survival and how complications were presented typified the knowledge claims made from clinical trials. On the one hand, the measure of success—the unit for comparing one treatment to another—was survival and its surrogate, disease-free survival. Almost the entire structure of the study, the whole of the statistical apparatus, was designed to ensure that the reported survival differences were significant and not a consequence of hidden bias. Sophisticated statistical methods were developed to address prerandomization, poststudy stratification, and a host of other methodological difficulties, always with the goal of rooting out bias in the reported survival. On the other hand, the analysis of toxicity had no such elaborate statistical paraphernalia to support it. Complications were presented as a stepchild of survival and characterized with qualitative terms like *minimal* and *acceptable*. This privileging of survival in the design and execution of clinical trials, however, provided a limited measure for translating trial results into local practice.

Although I do not discuss the details of Saenger's total-body irradiation program until chapter 4, it is worthwhile to briefly locate his studies within the clinical trial structure I have just outlined. His experiments were not randomized studies like Fisher's L-PAM trial but more exploratory and concentrated on toxicity and not on survival. Indeed, the military component meant that Saenger was effectively doing a dose escalation trial through the mid-1960s. Yet it is striking that, when he came under criticism in the early 1970s for the excessive toxicity of his program, he countered by producing stratified survival statistics of the kind that Fisher used. He claimed that he had improved the survival of many patients and that complications were comparable to those seen in treatment with chemotherapy. Although his studies were not multi-institutional, as Fisher's were, they shared the same methodologies and had similar difficulties. He had to recruit coinvestigators, clinicians, allied health workers, and others from a large network, and he had similar problems enrolling patients in the face of competing protocols. And, to maintain control of the practices of numerous health workers, he produced written protocols prescribing therapeutic procedures, patient entrance requirements, and measurable end points for studies that were meticulously documented in yearly reports. Saenger, however, shared more with Fisher and other postwar researchers than entrepreneurial acumen, written protocols, and reports and other paraphernalia that are so characteristic of clinical trials. He shared deeply in the clinical trial ethos that medical and scientific knowledge could be gained only, as Fisher claimed, through careful and controlled clinical trials.

Controversy and the Road to Closure

Even exemplary trials, such as Fisher's L-PAM study, which was meticulously designed and delivered and thoroughly analyzed, were subject to endless criticisms. The critics of clinical trials inevitably pointed to the numerous limitations in the design and execution of trials; for example, the patient groups were not appropriate proxies, or the clinical measures of success, such as disease-free survival, were not useful surrogates for long-term survival. Other criticisms pointed to more generic problems; for example, the small survival differences between the arms of a study make it difficult to determine a statistically significant difference without very large and long-term studies, or the trials are ethically wrong, and so on.[43]

The difficulty of bringing these endless arguments to closure became especially worrisome by the mid-1970s. The euphoria surrounding the 1971 Cancer Act had waned, and congressional critics of health care pressed the NIH to take the lead in finding a way to resolve the controversies surrounding the claims of clinical trials and the assessment of new medical technologies. Donald Fredrickson, who headed the NIH at that time, worried that, if the medical community could not replace its "informal but often haphazard process for creating authority by increment," then outsiders might set up "creations of 'technology management,' which may rely unduly on regulatory measures, or marketing controls."[44] Fredrickson feared that, if medicine's knowledge claims continued to be highly contested and the acceptance of new techniques remained so grudging, the medical community would relinquish some of its independence to government agencies. As a result, he created a special branch at the NIH[45] to facilitate consensus development.[46] This program became a model for later developments of technology assessment throughout much of Western Europe. In particular, the NCI consensus program that began in the late 1970s consisted of highly structured two and a half day proceedings at which an invited panel of experts produced a consensus statement after hearing presentations by scientists and other advocates as well as comments from an open forum.[47] This consensus process was, in keeping with the ethos of unbiased clinical trials, presumed to arise out of a "commitment to basing consensus strictly on examination of evidence by a neutral panel."[48]

In spite of the introduction of consensus conferences, knowledge claims based on clinical trials often remained highly contested. To give an appreciation of these difficulties, in this section I follow a particular example — the controversies surrounding whether chemotherapy should be given as an adjunct to mastectomy (or as an adjunct to local excision and radiotherapy) in treating node-negative breast cancer patients.[49] Indeed, the current standards of practice were the result of a very contentious series of debates that took place between 1980 and 1992. In 1980, an NCI consensus panel reviewed the role of chemotherapy in breast cancer and did not recommend adjuvant treatment for node-negative patients, even though the panelists were aware that approximately 30 percent of the patients would subsequently develop metastases. By 1992, the St. Gallen's Conference, a meeting of medical, surgical, and radiation oncologists, recommended that virtually all node-negative patients should receive some form of adjuvant treatment.[50] This conclusion was reached in spite of the fact that, during the intervening years, no clinical trial had demonstrated increased *survival* with adjuvant therapy. How, then, did we get from the 1980 NCI consensus conference with no adjuvant therapy to textbook decision trees with most of their branches ending in chemotherapy or hormonal therapy?

To be sure, there were a number of studies beginning in the early 1980s that demonstrated increased *disease-free* survival in breast cancer patients receiving various types of chemotherapy or hormonal agents.[51] The randomized trials were carried out in a number of countries, and five of those trials, identified as Milan IV, NASBP B-13, Intergroup 0011, and Ludwig V, are often quoted as the most important. Each of the trials differed in a number of respects: for example, they used several two- or three-agent chemotherapy regimes; and their patient populations varied from ninety in one trial to almost thirteen hundred patients in another. The NASBP B-14 trial stood alone since it used a hormonal agent (tamoxofen) in over twenty-eight hundred patients.[52] It was not possible to cite the results of any of these studies without raising new debates about the differing protocols, the patient entrance requirements, the number of patients, the trial location, and especially whether disease-free survival was a legitimate proxy for survival.[53]

A second NCI conference in 1985 also recommended against adjuvant therapy.[54] Since the clinical trials Milan IV and Ludwig V had ended by 1985, early results on the efficacy of adjuvant therapy were available, as was a meta-analysis (see below) of extant trials, but this was still not sufficient to swing the tide toward chemotherapy.[55] By July 1988, the debate was ratcheted up when the NCI issued a clinical alert regarding three American trials that reported increased *disease-free* survival with adjuvant therapy from 5 to 15 percent at approximately four years. The head of the NCI, Vincent DeVita, argued that these findings were too significant for patient treatments to be held hostage to the completion of peer review since "the hormonal and chemotherapy treatments described represent credible options worthy of careful attention."[56] To obtain the approval of the investigators for early release of their results, DeVita had convinced the editor of the *New England Journal of Medicine* not to enforce the rule prohibiting the publication of studies previously released to the news media.[57] The *Journal* later published the three studies with commentaries by W. L. McGuire and DeVita and a number of letters to the editor.

McGuire and DeVita debated two issues: the appropriate way to measure therapeutic efficacy and the true subjects of the trials. McGuire aligned himself with the 70 percent of the patients who would not recur but who would, nevertheless, receive toxic chemotherapy. His measure of efficacy was, not survival alone, but a broader indicator that addressed questions about toxicity and the long-term effects of treatment and its costs.[58] DeVita, on the other hand, focused on the 30 percent of the patients who would recur without adjuvant chemotherapy.[59] According to him, toxicity and cost were minor issues compared to the problem of tumor recurrence—it was survival, not complications, that mattered. DeVita argued that while "toxicity and the effect of these treatments on fertility . . . are of concern . . . so is the morbid effect of a recurrence . . . and the 'toxicity' of a premature death."[60] Throughout the clinical alert and the subsequent arguments, DeVita's aims were to narrow the debate and force closure. But he unleashed a storm of controversy that only raised new debates. Nevertheless, the early release of clinical trial results forced the opponents into a defensive posture and put considerable pressure on physicians, who became concerned about the medical/legal implications of not following the NCI alert.[61] Many physicians were also upset that the NCI was trying to control their clinical decisions—which, in essence, it was; the alert says as much.

The NCI held yet another consensus conference in 1990, in part to try to settle the controversy surrounding the clinical alert. The conference addressed four issues: the role of mastectomy versus lumpectomy combined with radiation therapy in early stage breast cancer, the optimal radiation technique, the role of adjuvant therapy in node-negative patients, and the role of prognostic factors.[62] The conclusions of the conference were important for two reasons. First, the combination of lumpectomy and radiation was deemed superior to mastectomy and represented a victory for radiation oncologists after the long and fearsome battle waged with surgeons for almost two decades. Second, the recommendation of the clinical alert—that adjuvant therapy should be seriously considered in node-negative patients—was reinforced. The conference was, however, criticized for not having been definitive enough. A June 1990 *Washington Post* article stated that the "panel of experts offers little guidance to women."[63] But that position underrated how much the conference had swung the debate toward the adjuvant therapy camp. The recommendation that chemotherapy should not be given in "patients with tumors 1 cm or less" was by default an endorsement of adjuvant therapy in all other patients who were at enough risk that "the decision to use adjuvant treatment should come after a thorough discussion with the patient."[64] In fact, the group of patients with tumors of one centimeter had never been a point of contention even for the strongest advocates of adjuvant treatment. They believed, not only that those patients were at low risk, but also that estrogen receptors—an indicator for using chemotherapy—could not even be measured in such small lesions. The 1992 St. Gallen's conference finally issued the positive version of the NCI consensus, namely, that adjuvant therapy should be given to everyone except for those patients in the low-risk category (essentially patients with tumors one centimeter or less).[65]

Since the early 1990s consensus statements were based on trial results similar to those available by the mid-1980s, why did it take so long to reach closure? To begin with, a data synthesis, a so-called meta-analysis by an international ad hoc group of medical oncologists, was able to demonstrate that, in the broadest sense, adjuvant therapy in the treatment of breast cancer led to improved survival.[66] The ethos underlying the analysis was similar to that of clinical trials; namely, it was a comparison of treatments using measurable clinical end points, statistical analysis, and scrupulous removal of all known bias. The analysis was carried out under well-defined protocols; for example, for a clinical trial to be entered into the meta-analysis, it had to meet strict entrance requirements, especially that the trial provided an unbiased estimate of the study questions. Once the entrance criteria were met, the trial was closed to further scrutiny and became a single data point in the meta-analysis, identified by an epithet like Milan-IV. All the details of the individual trials, all the messiness, the imperfections, and the battles over their worthiness, were put to rest once the trial became a valid member of the metaworld of clinical trials. That world was, however, markedly different from the one that existed prior to the meta-analysis. To begin with, all nonrandomized trials were excluded, which meant that a significant majority of the trials that had been used in prior debates were no longer in the running. Second, unpublished randomized trials (invariably those that show no difference between treatment arms) were meticulously sought out and included in the study to eliminate the effect of publishing bias.

Meta-analysis was like an idealized consensus conference, where all the contestants were at the table, each question was represented by a quantitative result like survival, and consensus was achieved when the difference in survival for competing therapies rose to the level of statistical significance. But, to achieve this ideal, the meta-analysis by necessity had to greatly simplify the clinical questions. The only measures of outcome that were acceptable were those "endpoints that are objectively measurable, such as overall survival," while toxicity, quality of life, and similar outcomes were explicitly excluded from the analysis since they differed "from study to study depending upon the mechanisms for follow-up and patient assessment."[67] A further simplification was the homogenization of the various patient groups, which was necessary to provide enough data to yield highly significant results.[68] The answers from the meta-analysis were definitive about survival but vague about which chemotherapy was best and which patient subgroups would benefit.

Because of the simplifications, the meta-analyses for node-negative patients could not be used to effect closure anywhere except in the idealized metaworld. Nevertheless, the statistical power of the results and the scrupulous attention to removing bias carried sufficient authority to close off certain avenues of debate. No longer could opponents of adjuvant chemotherapy in node-negative patients contest its value by arguing that it had never been conclusively shown to be effective in any breast cancer patients. Although meta-analysis could not identify which subgroups of node-negative patients would benefit, the debate had shifted to when and in whom, rather than whether, to give chemotherapy.

Although the meta-analysis provided one more constraint to the debates, it could not of itself force closure. It appears that other issues, including social ones, finally led to closure in the early 1990s. If nothing else, the pressure on oncology researchers to give definitive guidance to women was very strong by the early 1990s. The arguments for and against chemotherapy in breast cancer had been going on since the 1970s and continued to undermine the image of the medical community. And there is little doubt that continuing research support depended on the medical community demonstrating success in certain critical areas—especially breast cancer therapy. Moreover, DeVita had raised the stakes with the clinical alert. On the one hand, the alert was meant to demonstrate to congressional critics that chemotherapy had within its grasp one more "cancer victory," and, on the other, hand, it placed additional pressure on the cancer researchers to reach a definitive consensus. Perhaps, too, the design of the 1990 NIH consensus conference may have encouraged a quid pro quo between the chemotherapy and radiation oncology communities. The recommendation that radiation combined with local excision should be used in preference to mastectomy alone in early breast cancer was a major victory for the radiation oncologists. Although the gains for chemotherapy were not as clear-cut, the recommendations of the clinical alert were reiterated, and the role of prognostic indicators, which had been used as an argument against chemotherapy, was defused by recommending further research. The political stakes at a consensus conference were best put by DeVita: "The disproof of a therapeutic hypothesis may mean the shift in management of an entire disease from one medical specialty to another. This change is generally not well received in medicine. Few clinicians appear willing to design their specialty out of a clinical experiment."[69] The knowledge claims on behalf of adjuvant chemotherapy were, as we have seen, not the necessary result of well-designed, unbiased clinical trials. Epistemic, social, and political issues all contributed to the consensus that was eventually reached.

Once consensus was finally attained among medical researchers, a formidable problem still remained, namely, convincing the clinical community to adopt new standards of practice. At first, researchers thought that the difficulty of changing local behavior was simply one of poor dissemination of the standards reached in their consensus conferences. In time, however, they began to appreciate the difficulty of introducing new standards into the clinical community and the variability of their influence for changing local practices. A further discussion of this topic would lead to a long digression.[70] Suffice it to say that the medical investigators were encountering an age-old problem of one community imposing standards on another. These problems are discussed in detail in Witold Kula's *Measures and Men*, which follows, from the Middles Ages through postrevolutionary France and Western Europe, the political and social implications of applying new standards, for example, those for measuring the quantity of wheat or the size of a plot of land. The efforts of local communities to resist the imposition of new measures, which they feared would have important economic and social implications, look and feel very much like the responses of postwar American clinicians to the claims of new practices announced by researchers following consensus conferences. Perhaps E. P. Thompson put it best in his discussion of the response of local communities to new standards (in his case, during the early modern period in England): "Attempts to change the measure often encountered resistance, occasionally riot."[71]

To return to the postwar landscape of clinical trials, we find it covered with larger and ever more carefully designed studies. As the trials grew in size and number, and as they influenced ever-larger cohorts of individuals, the battles over the epistemic claims of clinical trials only increased. To counter these arguments, new institutions, cooperative groups, and more sophisticated methods of analysis arose. But the widening scope of the medical and lay communities affected by clinical trials and by new institutions and cooperative groups further increased the number and range of the grounds on which medical knowledge could be contested. For clinical trials, the knowledge claims were not contested only on epistemological, social, and political grounds. With the growing postwar concerns with clinical conduct, medical investigators also had to demonstrate that their research was produced ethically. Eugene Saenger's studies need to be located within this postwar clinical trials environment. Those studies, which covered the decade of the 1960s, were carried out during a period when the various clinical trial methodologies were beginning to reach maturity. At the same time, ethical regulation was not at all stable and went through major changes as the NIH implemented, in the mid-1960s, a program of local peer review that included the introduction of informed consent. I turn in the next chapter to this volatile and changing environment of ethical regulation.

Notes

1. Fisher, "Clinical Trials for Cancer," 2609.

2. Other cancer specialties followed the chemotherapists, e.g., Simon Kramer's work on creating a network of multicenter trials in radiation oncology (Kramer, "Radiation Therapy Oncology Group"). For a history of the development of the Radiation Therapy Oncology Group, see Cox, "Evolution and Accomplishments."

3. For example, the MRC had established the curative role of streptomycin in tuberculosis, while, in a less well-known but equally definitive trial, it concluded that antihistamines had no influence on the common cold (Medical Research Council, "Streptomycin Treatment of Pulmonary Tuberculosis," and "Clinical Trials of Antihistaminic Drugs").

4. Fisher, "Clinical Trials for Cancer," 2611.

5. For a particularly cogent presentation of the controversies over clinical trials, see Lawrence, "Some Problems with Clinical Trials," 373–78. See also Meinert, *Clinical Trials*, 17–18.

6. Zubrod, "Clinical Trials in Cancer Patients," 185.

7. Lasagna, "Controlled Clinical Trial."

8. Zubrod, "Origins and Development," 11.

9. Lasker later played a significant role in the Cancer Act of 1971 (the War on Cancer). For the War on Cancer, see Strickland, *Politics, Science and Dread Disease*, 288–91; and Rettig, *Cancer Crusade*, 281–315.

10. Zubrod et al., "Historical Background," 10. See also Keating and Cambrosio, "From Screening."

11. Zubrod, "Origins and Development," 12.

12. Laszlo, Cure of Childhood Leukemia, 91.

13. Löwy's argument follows Theodore Porter's contention (in *Trust in Numbers*) that statistics are often taken up by weak disciplines to ward off the efforts of outside groups (*Between Bench and Bedside*, 52).

14. Sullivan, "Placebo Controls," 218.

15. Marks, *Progress of Experiment*, 10.

16. For example, during the war, Zubrod worked on clinical trials to treat malaria for the military. He recalled: "I learned lessons about how to organize clinical trials and those lessons served me well later on when I organized a group to cure leukemia" (quoted in Laszlo, *Cure of Childhood Leukemia*, 97).

17. Ibid., 100.

18. Freireich and Lemak, *Milestones in Leukemia Research*, 76.

19. As early as 1957, chemotherapy consumed half the NCI budget (Patterson, *Dread Disease*, 196).

20. DeVita, "Evolution of Therapeutic Research," 908.

21. Fisher et al., "L-Phenylalanine Mustard," 2899.

22. Ibid., 2884.

23. Fisher, "Laboratory and Clinical Research," 3867.

24. Hill, "Clinical Trial" (1951), 280.

25. In this respect, the L-PAM study was unusual since, in most cancer clinical trials, it is not possible to do double-blind studies.

26. Patterson, "Clinical Trials in Malignant Disease," 80; Fisher et al., "L-Phenylalanine Mustard," 2885.

27. Byerly, "Explaining and Exploiting Placebo Effects," 425, quoted in Sullivan, "Placebo Controls," 221. To be more specific, blinding the patient in the study does not remove the placebo effect but simply distributes placebo cures over both arms of the trial. Placebo cures are noise, contaminating the trial, and diminishing the statistical power of the results (Sullivan, "Placebo Controls," 221). In the period during and following World War II, the efficacy of placebos was generally recognized. Although many in the medical community dismissed the placebo effect as merely a psychological artifact, Henry Beecher, a well-known investigator in physiology and a fierce critic of unethical practices in clinical trials (see chapter 2), took another position. His research at Harvard led him to reject the notion that "the action of placebos is limited to 'psychological' responses. Many examples could be given of 'physiological' change, objective change, produced by placebos" (Beecher, "Powerful Placebo," 1603, quoted in Sullivan, "Placebo Controls," 222). Because of his belief in the reality of the placebo effect, Beecher championed double-blind studies and was concerned with formal ethical regulations when they would jeopardize the therapeutic role of the doctor-patient relationship (Beecher, "Experimentation in Man"). See also Shapiro and Shapiro, *Powerful Placebo*.

28. Because of this, Hill argues: "A trial should be begun at the earliest opportunity, before there is inconclusive though suggestive evidence of the value of the treatment" ("Clinical Trial" [1951], 279).

29. Gehan and Lemak (*Statistics in Medical Research*, 144) argue that clinical equipoise, in which the arms are in balance for an individual patient, is very rare

and that the promise of the new therapy tested in the trial is what provides the ethical warrant for entering patients. See also chapter 2.

30. A declaration of death eliminates all possibility of bias since, as Hill quipped, "no statistician, as far as I know, has in this respect accused the physician of an over-reliance upon clinical impression" ("Clinical Trial" [1952], 117).

31. Fisher's own was the National Surgical Adjuvant Breast and Bowel Project.

32. Fisher was cleverly skirting the standard sequence, at least in the textbooks, that dose escalation should be carried out in Phase I studies prior to the initiation of Phase III randomized trials (Pocock, *Clinical Trials*, chap. 1).

33. Meinert, *Clinical Trials*, chap. 5.

34. Cox, "Brief History," 27. For centers of calculation, see Latour, *Science in Action*, chap. 6.

35. Fisher, "Clinical Trials for Cancer," 2613. For the early randomization procedures used in the United Kingdom, each new patient was assigned a number in sequence, and an envelope marked with that number was opened to determine the patient's assigned treatment (Hill, "Clinical Trial" [1951], 280). The meticulousness with which Fisher conducted the multicenter trials led him to drop a similar method in favor of centralized randomization since "the possibility exists that the contents of the envelope may become known before patient entry into the trial" ("Clinical Trials for Cancer," 2613).

36. Hellman and Hellman, "Of Mice but Not Men," 1586.

37. Meinert, *Clinical Trials*, chap. 5.

38. Fisher, "Clinical Trials for Cancer," 2614.

39. Begg et al., "Participation of Community Hospital Clinical Trials," 1076.

40. Fisher et al., "L-Phenylalanine Mustard," 2889.

41. Ibid., 2897, table 7.

42. Ibid., 2899.

43. Lawrence, "Workshop on Clinical Trials," 373–78.

44. Fredrickson, "Seeking Technical Consensus," 116.

45. The section was known as the Office of Medical Applications for Research.

46. Perry and Kaberer, "NIH Consensus-Development Program," 169.

47. Jacoby, "Consensus Development Program," 108.

48. Jacoby, "Update on Assessment Activities," 3039. The belief that rational discussion by experts would lead to uncontested truth claims also produced attempts to reach closure through the improbable idea of a science court of experts (Task Force of the Presidential Advisory Group, "Science Court Experiment," 653–56).

49. Abeloff et al., "Breast," 1673.

50. Glick et al., "Adjuvant Therapy," 296–97.

51. Actually, of the many trials that started in the early 1980s, only the so-called Milan–IV trial showed a survival advantage, but that result was universally questioned because of the low survival rate of the control arm (Abeloff et al., "Breast," 1671).

52. Ibid., 1672.

53. The standard argument is that an increase in disease-free survival does not imply a similar improvement in long-term survival since the former may indicate only that

recurrences will appear later. In the case of breast cancer, ten-year survival is the sine qua non of success (at least it was in the period under discussion here), and the trial results in the late 1980s did not and could not answer the long-term survival question.

54. NIH Consensus Conference, "Adjuvant Chemotherapy," 3463.

55. "Review of Mortality Results," 1205.

56. Omura, "Clinical Cancer Alerts," 423.

57. Ibid., 424.

58. McGuire, "Adjuvant Therapy," 526.

59. DeVita, "Breast Cancer Therapy," 528.

60. DeVita, Letter to the Editor, 472.

61. Prosnitz, "Medical and Legal Implications," 1574.

62. NIH Consensus Conference, "Early Stage Breast Cancer," 392–94.

63. Omura, "Clinical Cancer Alerts," 428.

64. NIH Consensus Conference, "Early Stage Breast Cancer," 394.

65. Glick et al., "Adjuvant Therapy," 296.

66. For an overview of the various Early Breast Cancer Trialists' Collaborative Group (EBCT) meta-analyses, see Gelber and Goldhirsch, "Overview to the Patient."

67. Ibid., 170.

68. A 1988 meta-analysis, e.g., combined twenty tamoxofen trials with forty chemotherapy trials and concluded that tamoxofen reduced the odds of recurrence by one-fifth in women over fifty and that chemotherapy reduced the odds by one-fourth in women under fifty (EBCT, "Effects of Adjuvant Tamoxofen," 1681).

69. A consensus conference decision may be difficult to reach if any of the parties believe that they have much to lose (DeVita, "Evolution of Therapeutic Research," 910).

70. The difficulty of transferring standards of practice in medicine is developed in Kutcher, "Cancer Clinical Trials."

71. Thompson, "Moral Economy," 217.

2 The Production of Trustworthy Knowledge

A refreshingly unstructured pronouncement emerged from some members of the French Academy of Political and Moral Sciences. Discussing the Nazi experiments they suggested that no written law can provide a strict ruling that knowledge gained for science constitutes the ultimate justification, but that the responsible, experienced, prudent scientist—who is the only type of experimentalist to be tolerated—will draw his own limit. This view is one that is certainly close to the heart of many scientists. • Louis Lasagna, *The Doctors' Dilemmas*

Medical research during the postwar period grew rapidly in power and prestige. Yet there was a fundamental dilemma at its heart. On the one hand, medical research practices, particularly randomized clinical trials, were highly regulated. Investigators increasingly had to adhere to strict standards of practice if their studies were to garner government funding and obtain acceptance within the research community. Indeed, in the previous chapter, we observed the highly regimented character of Fisher's studies—and they were by no means unusual. On the other hand, medical research did not maintain a similar system for the regulation of clinical conduct. Researchers believed that ethical behavior could continue to be maintained through the virtue of each physician-investigator. The epigraph to this chapter, by Louis Lasagna, one of the most thoughtful of the postwar investigators, and someone who wrote often on matters of research ethics, makes evident the strongly held belief within the medical community that research decisions should be left in the hands of virtuous physician-investigators.

As more and more research money entered the arena in the 1950s, and as an increasing number of investigators participated in clinical studies, the medical research enterprise began to falter. Public disclosures of questionable and sometimes outrageous conduct by researchers began to fill the news media. At the same time, physicians came increasingly to fear legal suits. There was no practical code of ethics that American researchers could use as a standard of practice, nor was there any history in case law that would provide them with protection from legal suits for causing harm with their experiments. In spite of these concerns, the American medical community seemed reluctant to produce a code of practice to govern medical research. There were a number of reasons why, and I discuss them in greater detail later in the chapter. But most significant was that physicians fiercely tried to hold on to the patient-physician relationship and its clinical prerogatives, which a more formal system of governance would, they believed, undermine.

A resolution to the dilemma came from outside the community of medical researchers per se when, in the mid-1960s, the National Institutes of Health (NIH) (and the Food and Drug Administration [FDA]) instituted formal rules for the regulation of medical research. These moves were precipitated by a series of highly publicized scandals about unethical practices that reached a crescendo in the early 1960s. Regulators, particularly at the NIH, began to realize that it was no longer possible for trust in the ethical probity of research to be held by investigators. Consequently, the NIH put in place a new system of governance that included a review by peers. In order to receive government funding, investigators had to demonstrate that they had passed a local ethical review that judged the form and content of a consent statement and the risks and benefits of the proposed research. This program transferred the regulation of clinical ethics from the judgment of individual investigators to medical/government institutions (local peer review committees). This system would restore the public's trust in research through demonstrable institutional practices that would control medical investigations. Indeed, chapter 6 shows how the local peer review committee at the University of Cincinnati constrained and, thus, influenced the direction of Saenger's research program as it tried to mediate between local institutional needs and government regulations.

My emphasis on the importance of the NIH system of regulation and my claim that it represented a radical change in the location of trust—from physician-investigators to medical/government institutions—differs from that of more traditional interpretations of the development of postwar medical ethics. To many critics and interpreters, the importance of the NIH system was that it signified one more step in the evolution of human autonomy (and its surrogate, informed consent) that began with the Nuremberg Code of 1947. The code, which was written following the so-called Doctors Trial in Germany, had as its first of ten principles that the "voluntary consent of the subject is absolutely essential."[1] As one historian saw it, the code "served as a blueprint for today's principles that ensure the rights of subjects in medical research."[2] For another, it was a "template for international directives on human research ethics."[3] In many ethically based accounts, the Nuremberg principles were reiterated and developed as a series of international codes that sought to elucidate and purify medical research, first to protect patients from harm, and later to protect the principle of human autonomy.[4] Indeed, to some contemporary bioethicists, the Nuremberg principle of informed consent is of such transcendent importance that any attempt to modify it, as in later international codes, is viewed with alarm and disdain.[5] Other interpreters of postwar ethics locate the fulcrum of change in a 1966 *New England Journal of Medicine* article by the medical researcher Henry Beecher that revealed twenty-two cases of unethical experimentation. Beecher's exposé is identified as the signal event leading to the introduction of ethical regulations, including those of the NIH and the FDA.[6] According to David Rothman: "The events in and around 1966 accomplished what the Nuremberg trial had not: to move medical experimentation into the public domain."[7] These events marked, in this interpretation, the beginning of the death knell of medical paternalism and led to the increasing importance of outsiders (strangers) in the regulation of research.

As I show below, the NIH system of governance was primarily about, not the growth of human autonomy, but, rather, reducing the level of risk in medical experimentation. Nor did the regulatory system that the NIH introduced mark the end of medical paternalism; rather, it represented the last in a series of changes that transferred authority from individuals to institutions. For scientific research, those changes had begun as far back as the early modern period. As Stephen Shapin argues, the premodern period, with its face-to-face encounters, located veracity in "gentlemen" whose free action and integrity were "seen as the condition for truth-telling, while constraint and need were recognized as the grounds for mendacity." In the modern period, knowledge and trust came to reside, not in the actions of free and unconfined individuals, but in "institutions which must vigilantly constrain the free actions of their members."[8] In like manner, the NIH expanded institutional review of scientific proposals, a common feature of postwar medical research, to include the governance of clinical conduct. The medical/government community would constrain not only scientific practices but also the ethical behavior of individual physician-investigators, thereby maintaining the trust of the public.

In the first section of this chapter, I show that Nuremberg played distinctive, but different, roles from that described above. First, in the immediate postwar period, the trial forced some in the American medical community to confront ethical issues of mass experimentation, if only to distinguish German medicine from American. Second, the Nuremberg Code entered into the medical ethical arena through the back door in that its consent requirement began to be used in negligence lawsuits and placed researchers at increased legal risk. The need for a workable code of practice became urgent. In the second section, I consider reasons why American medical researchers were unable to produce a system of governance, in spite of their looming legal problems as well as increasing public concern about unethical experiments and a growing loss of trust in the medical community. In the last section, I discuss the NIH's system of governance and how it provided a framework within which the medical research enterprise could control its practices and continue to flourish.

In this chapter, I place special emphasis on the political stakes and the pragmatic individual and programmatic needs of those in the medical community. It is important to understand the needs of these researchers as they tried to negotiate the many hurdles standing between their nascent intentions and successful research programs. Without such an analysis, from the ground up, it would be difficult to fully appreciate both Saenger's efforts and those of the peer review committees that attempted to control his program. One could also investigate ethical regulation from the top down, viewing clinical conduct in light of moral arguments and, particularly, bioethical principles.[9] But such an approach would produce a very different understanding of postwar practices, which were, I believe, primarily governed by concerns about how to realize particular goals (e.g., launching and maintaining research programs) within the ambient political and social culture. In these terms, ethical review committees and informed consent statements were but two issues among the many that researchers had to confront.

Postwar Clinical Ethics Discourses

The period between the end of World War II and the middle of the 1960s, between the Nuremberg Code and the regulatory structure put in place by the NIH, should not be considered as an ethical wasteland where there was little concern with ethics except, perhaps, for occasional hand-wringing following the effluence of a scandalous experiment. If we read the literature, especially that in the medical journals, we find a flurry of papers and conference proceedings on medical ethics. It seems, at times, at least for a coterie of physicians, as if there was little else they could or wanted to write about. If there was little consensus about what type of ethics was appropriate for the new medical research environment, there was little doubt that this coterie of physicians believed that there were ethical problems that needed solving. To Henry Beecher, "the purposes of human experimentation have become deeper and more complex," and, as a consequence, so had "the problems surrounding it."[10]

The ethical self-scrutiny of at least some of the medical community had limited wartime or prewar precedents.[11] Although most American physicians ignored, as the argument goes, the Nuremberg Code as one for "barbarians," that should not mislead us into believing that it had no immediate influence on the ethical considerations of any of them. The trials in Germany forced some in the U.S. medical community to appreciate the possible excesses that could occur as a consequence of the growing importance of large-scale medical experimentation. And, if American physicians dismissed the activities of German physician-researchers as beyond the standards of Western conduct, they were still forced to consider their own ethics, if only to distinguish themselves from the German physicians.

In fact, the American medical community had to deal with the German practices immediately following the war. In the spring of 1946, Andrew Ivy, an American physician and researcher, went to Nuremberg at the request of the secretary of the army to serve as a medical expert in the Doctors Trial (*United States v. Karl Brandt*). Ivy was chosen by the American Medical Association (AMA) following a request by the army surgeon general for a medical expert for the upcoming trial of German physicians accused of crimes against humanity. The choice appeared apt. Ivy had excellent credentials as a medical researcher and had served as a civilian director at the Naval Medical Research Institute in Bethesda. Early in the war, he had supervised medical experiments for the American war effort, experiments that included seawater desalination studies with human volunteers as well as physiological studies in aviation medicine. Similar experiments had been included in the indictment against the German medical researchers.[12]

After Ivy's return from his first visit to Nuremberg, where he had found a state of confusion in the War Crimes Branch over the ethical and legal aspects of human experimentation, he presented his ideas on ethical practices to the AMA's Board of Trustees. The AMA leadership had no doubt that the German experiments were a gross violation of standards that they, the leadership, believed were already inherent in the existing AMA Principles of Medical Ethics.[13] Those general standards were no different than those found in the AMA Code of Ethics of 1847, which expressed an ethos of scientific advancement for the benefit of humanity: "To expound medical science, and to extend its domain of practical application and usefulness, . . . [both] must be the product of a regular and comprehensive system."[14] The code, however, contained little in the way of practical and direct guidance regarding human experimentation. In recognition of the absence of specific ethical guidelines, the AMA rapidly took up Ivy's proposed rules of conduct. Ivy submitted a written report in September 1946, and, by December, the report had passed through a number of the AMA's committees, where it was modified and distilled into the three ethical principles that were added to the AMA Code. Experimentation on humans required voluntary consent, it also required previous tests on animal systems, and it had to be performed under proper medical protection and management, that is, under the direction of physicians.[15]

The rapidity with which the new code moved through the AMA hierarchy was astonishing, and it illustrates the desire of the AMA to distance itself and the American medical profession from their German counterparts. A code of practice with specific rules, which was, presumably, part of a long tradition in American medicine, would bolster American medical prestige and provide standards to set against the Nazi experiences. The code with its specific principles was also critical ammunition for Ivy to take back to Nuremberg. The trial began (on December 10) just one day prior to the approval of the new AMA Code. At the trial, under direct examination, Ivy claimed that his principles represented those of the AMA.[16] He failed, however, to mention under direct questioning what he was forced to admit under cross-examination, namely, that the principles in the AMA Code had only recently been approved.[17]

The lack of debate at the AMA surrounding the new ethical principles also illustrates early cold war thinking about the character of medical research. American physicians identified research in Germany with mass experimentation using putatively healthy subjects, and they distinguished those studies from those of American physicians who practiced small-scale therapeutic research on patients. The latter type of study was viewed as part and parcel of a long tradition of medical care under which physicians modified a treatment regime to reflect the patient's unique and special circumstances. Such maneuvers were believed to be conservative and to not veer overly much from tradition and local practice. In contrast, Ivy's first principle (on consent) appeared to be aimed more at the use of healthy volunteers in experiments that were for society's gain, experiments that certainly required consent. In fact, there was a steady tradition of consent in experiments with healthy volunteers. To take one example, Walter Reed and his colleagues in Cuba deliberately infected human subjects (including some members of the research team) with yellow fever, but only after giving the volunteers written contracts the terms of which included monetary compensation and any necessary medical treatment.[18] It is likely that the AMA hierarchy, as well as Ivy himself, considered the new code primarily, if not entirely, as a prohibition on this voluntary form of medical experimentation. Therapeutic experiments fell into a different ethical category since they were covered by the physician's fiduciary responsibility to the patient. The Nuremberg court itself followed a similar line of reasoning since the indictments were for crimes against putatively healthy subjects. The AMA Code, like the Nuremberg Code, addressed experiments on healthy volunteers.

The sequestering of therapeutic research under the umbrella of the virtuous behavior of physicians and the distinguishing of it from mass experimentation with healthy volunteers was reiterated in the early cold war ethical discourses of a coterie of American physicians and researchers, including Welt, Wiggers, Shimkin, Lasagna, Beecher, Guttentag, and others. These names appear over and over again in papers and conferences on medical ethics in the 1950s and 1960s. It is not clear whether other researchers who did not write about medical ethics shared similar concerns, but it is likely that they did. Most had been educated during or shortly following the war, with the result that their attitudes toward the importance of research, the tradition of experimentation by physicians, and the extreme character of the German experience should have been similar to those of high-profile researchers.

The concerns of these physician-ethicists tended to follow along the lines of Ivy and the AMA. Healthy research subjects were associated with large-scale studies that were outside the physician-patient relationship and, thus, in need of the protection of informed consent. Research on sick patients, however, was viewed in the glow of a "simpler" past where physicians experimented on individual patients who were protected by the patient-physician relationship. For example, the medical researcher Michael Shimkin wrote in a 1953 article: "Medical experimentation on human beings in the broadest meaning and for the good of the individual patient, takes place continually in the doctor's office." Clinical trials, Shimkin argued, were really no different: "The general question of human experimentation is one of degree rather than kind. Deliberate experimentation on a group of cases with adequate controls rather than on individual patients is merely an efficient and convenient means of collecting and interpreting data that would otherwise be dispersed and inaccessible."[19] This vision of clinical trials as little more than an extension of small-scale research under the guidance of physicians whose only interest was in the benefits to the individual patient was not unusual. One commentator noted that anything less amounted to a coercion of patients for experimental purposes and would not be "justified on a single person, even if millions of other lives could be saved by such an act."[20] Moreover, since therapeutic experiments were, presumably, incremental and fashioned for the individual patient, the likelihood of untoward consequences was minimal. For example, one physician informed a graduating class of medical students that their patients should not fear medical research since such concerns were "engendered through rumors or suspicions that experimentation involves discomfort, impairment of health, injury to the body, and perhaps some risk of fatality. Such misgivings are wholly figments of the imagination, they are not true in fact."[21]

This view of patients and their role in therapeutic experiments would rapidly change. Although clinical trials began shortly following the war, by the 1950s the number of large-scale studies burgeoned, and investigators began to understand the importance of recruiting sick patients to carry out their research. For example, Michael Shimkin, who viewed patients as part of small-scale experimentation, nevertheless also realized that hopelessly ill patients were an unparalleled opportunity for larger studies in humans. Shimkin declared that it was "the experience of many physicians that this type of patient often wants, even demands, that something be done for the advancement of knowledge, if not personal benefit."[22] Patients had become a necessary resource, one that had to be nurtured. Henry Beecher, writing as late as 1959, commented on "the rather newly recognized fact that some types of basic scientific advance can be made only in the presence of disease."[23] In the same year, a report from the National Conference on the Legal Environment of Medical Science concluded: "Evaluation of most drugs, devices and techniques may be carried out completely within the physician-patient relationship as generally understood."[24] The conference had met to discuss the increasing concerns with legal problems of experimentation, to which I return to below, but the report clearly equated experimentation with clinical research with, not healthy volunteers, but sick patients.

In the midst of this expansion of clinical trials and the growing recruitment of patient subjects, the Nuremberg Code became of concern to a number of investigators. There were various aspects of the code that critics considered a threat to the vitality and development of medical research. Although some of them will be discussed in the following section, the most pressing issue for physicians was the increasing fear that codes could be wielded in legal cases brought against practicing physicians. There was detailed criticism of the Nuremberg Code at the 1959 National Conference on the Legal Environment. The code's first principle (of informed consent) was of particular concern because of its "many legal technicalities which might leave research workers open to unrealistic damage charges."[25] Such concerns were not new and had only increased over the decade of the 1950s. For example, one critic noted: "Since these [the Nuremberg rules] represent the only code developed by a legal tribunal it is assumed that they have some validity in a court of law today."[26] And Henry Beecher worried that clinical research, which was essential, had "led to an expansion of human experimentation within our ethical and moral concepts even though it is still considered by the courts as extralegal in character."[27]

Researchers were clearly concerned about their exposure to legal action, and there was little in the way of precedent and protections. The standard tome, *Doctor and Patient and the Law*, advised medical researchers to beware since "in the treatment of the patient there must be no experimentation."[28] Indeed, the only support for human experimentation to which physicians could turn was the Michigan Supreme Court's 1935 judgment in *Fortner v. Koch*, in which it was acknowledged that human medical experiments were important and possible. But the condition set down by the court of not deviating too radically from accepted practice was too restrictive for medical experimentation.[29]

While there was no developing case law that would protect physicians conducting medical research, there was at the same time a disturbing shift in the nature of malpractice suits. Attorneys began to sue physicians for negligence on the basis of the insufficiency of informed consent rather than on the claim that the defendant had deviated from the local standard of practice. To be sure, the use of consent in legal suits had a long history, but it had been reserved for battery torts. Now, with consent as a ground for legal suits based on negligence, attorneys could argue that a patient had suffered a complication as a result of a statistical probability and that the patient would have not agreed to the procedure if fully informed.[30] This shift to the use of consent in negligence suits occurred, in part, because attorneys found that such a strategy made it easier to win such cases; they no longer had to rely on physicians testifying against each other, as they had when they sued on the basis of a deviation from the local standards of practice. Exposure to negligence based on the adequacy of consent made investigators particularly vulnerable since all clinical trial research carried risks that were sometimes difficult to predict. To take one example, in the 1960 case of *Natanson v. Kline*, Mrs. Natanson sued the radiologist Dr. Kline for severe toxicity from cobalt therapy following mastectomy. The court found Kline guilty of negligence for failing to adequately inform the patient of the risks of the treatment.[31]

There was much discussion at the time in the medical press and the newspapers about medical malpractice, and physicians came to fear such cases.[32] Physicians and medical researchers had reason to be concerned about the direction such litigation had taken. There was no standard developed by the community that defined informed consent or the ethical practice of medical experiments to which researchers could point in court. The only standard was the Nuremberg Code, and the medical community considered its regulations, especially the unconditional demand for informed consent, too strict for effective research.

Henry Beecher argued that the medical community needed to fashion a more workable code than Nuremberg in response to the courts. He implored the World Medical Association (WMA) in 1960 to address the problem. He reminded physicians that "in the United States no rules, codes or regulations have been generally adopted." An exploration of this issue must be undertaken "since the crucial study of new techniques and agents must ultimately be carried out in man." He hoped that "some preliminary but very general code might be developed which would serve the best interests of science and mankind in the realm of human experimentation."[33] Beecher was quite likely referring to a draft code of the WMA's that would later become its Helsinki Declaration—the first attempt at setting ethical standards in medical research since the Nuremberg Code—but movement was slow, and there appeared to be no other formal code on the horizon. The community did not have an effective ethical framework to support research practices.

The Reluctance of Medical Researchers to Standardize Clinical Ethics

Why did medical researchers find it so difficult to articulate prescriptive research ethics? There were good reasons to look toward standardization in ethics since it appeared to work so well in controlling the medical procedures carried out by physicians on clinical trials. Indeed, modern medicine was based on a widespread metrology (a system of setting and propagating standards) that pervaded, not only research, but also the delivery of all medical care from the use of the household thermometer to the uniform measures for interpreting diagnostic tests and for prescribing and delivering treatments. Since standardization was so ubiquitous in routine scientific and clinical practice as well as in large-scale clinical trials, and since popular belief in the power of modern medicine was (at least in part) a product of standardization, then why not apply a similar approach to ethics? Such a metrology would provide a clear guide for practitioners and could be used by the courts to adjudicate malpractice suits. Rather than the uncertain legal situation that physicians faced in the 1950s, they could have produced a set of ethical standards for regulating research practices. Such a system would have provided them with some legal protection and would also have restored the trust of the lay community, a trust that was eroding as a result of claims of unethical practices.

For the most part, however, the medical community did not or could not embrace the notion of standardizing the ethics of research practice. It could not appreciate that trust in the ethical behavior of physicians might come from the regulated practices of the medical community as a whole rather than from the virtuous actions of individual physicians. Michael Shimkin put it best for the medical researchers of his day: "Science per se is neither moral or immoral; it becomes moral or immoral only as moral or immoral human beings use its powerful techniques."[34]

The medical community also did not consider developing a formal system of internal regulation because it was concerned that formal codes could be the thin end of a wedge that might be used by external bodies to control medicine. Instead, it sought to defend its prerogatives from the possibility of outside interference. For example, at the 1959 National Conference on the Legal Environment, the Committee on Legal Battery and Clinical Research (CLBCR) recommended: "No legislation is indicated to regulate investigative procedures because there are sufficient codes and ethical practice in the medical profession." Yet the committee understood that there were not sufficient codes and legal precedent to protect medical researchers since it warned physicians: "Many insurance policies will not cover intentional legal battery in an investigative procedure."[35] Their recommendation against additional legislation, however, is best understood in light of the fear that pervaded the medical community that supporting such legislation would bring further outside influence into medical matters, something that should be avoided even at some cost to the community.

Indeed, the medical community believed that it had a contract with society giving physicians an authority and independence that should not be broached. That contract, as specified in the 1847 AMA Code of Ethics and its progeny, had become ingrained in American cultural and social arrangements.[36] The code bound physicians, patients, and society through an agreement in which physicians would tend to the sick and dedicate themselves to the welfare of society. By the postwar period, society had been amply rewarded with substantial advances in treatment, especially of infectious diseases. In return, physicians expected and received material prosperity, the respect of the community, personal and professional authority, and freedom from interference by state institutions. They believed that this contract had no room for the intervention of the state into their clinical and research activities even in the name of protecting patients.[37]

The fear of outsiders also took on a particular cold war—almost paranoid—fear that any government intrusion was but the first step in a totalitarian wedge that would compromise and, ultimately, destroy medical research. This fear was powerfully conveyed in a number of Andrew Ivy's writings in which he claimed that German medicine had been corrupted by the state. He referred, for example, to the development of two types of typhus vaccine by the prisoners in Block 50 of Buchenwald. The prisoners provided their captors with a vaccine for the German troops that was ineffective and harmless while they reserved the active batches for themselves.[38] The ruse was possible, according to Ivy, "when persons having a nazi-SS [sic] type training were placed in charge of experimental work under conditions where the aim was to seek political preferment or to follow scientific orders rather than to seek the scientific truth."[39]

For Ivy, such examples demonstrated "how a political gang, how an ideology such as totalitarianism can insidiously take over a profession, which has high ideals, and finally subvert that profession." The American medical profession must be vigilant regarding state control; otherwise, it might follow the "German medical profession and German medical science," which had "permitted these things to happen."[40] The medical investigator Michael Shimkin expressed a related concern that research programs aligned to the needs of the state might "introduce the sort of thing that is called the duty of scientific man to society . . . under which all sorts of monstrosities have been practiced in absolutist states."[41] The incursion of the state into scientific practices was to be greatly feared. For Ivy, it would lead to direct corruption of research; for Shimkin, it would encourage research to meet the needs of the state that would lead to human abuse. Neither Ivy nor Shimkin actually discerned that they and other American researchers had participated in wartime research expressly to meet the needs of the state. Nor did either quite appreciate that American researchers continued their political involvement after the war through heavy government support. Shimkin's and Ivy's vision was that of Vannevar Bush in *Science, the Endless Frontier*, according to which researchers would decide how science operated while the state merely provided unencumbered funding.

American medical researchers were not only concerned that the state would use ethical codes to control, if not corrupt, open scientific inquiry; they also feared the effect that formal codes would have on the patient-physician relationship, one of their cherished and closely guarded prerogatives. Physicians understood that, with its requirement of informed consent, the Nuremberg Code conflicted with that relationship. The subtle, important tacit understanding between patient and physician could not be sustained in a highly regulated system. Not only would therapy be jeopardized, but so would research. Henry Beecher was emphatic on this point: "In most cases, the problems of human experimentation do not lend themselves to a series of rigid rules."[42] Other researchers also expressed concerns that ethical codes would adversely affect their research efforts. At the National Conference on the Legal Environment, the CLBCR warned the medical community: "A legal code would only give rise to so many interpretative complications as to possibly inhibit legitimate research."[43] Another physician warned researchers "how ambiguous a code can be and how a lack of precision could be devastating in many ways to the proper development of a research program." Yet another stated: "Despite the fact that this set of rules [Nuremberg Code] was carefully developed by able men, they . . . tend to be restrictive and ambiguous."[44]

Louis Welt, as well as other researchers, argued that formal codes would not necessarily protect patients since it was often "impossible either to evaluate the risks this precisely or to communicate with the subject in such a fashion that he fairly sees the problem in all the dimensions."[45] Beecher echoed the difficulty of meeting consent requirements: "How is the investigator to draw a practical line in the prior information to be given his patient, between 'reasonably expected' and 'possible hazards,' when these will be quite unknown in first experiments?"[46] The points were clear: codified ethical practices were impractical and would hamper medical research and fail to effectively protect patients. Even physicians would not necessarily gain protection. Beecher, who had previously exhorted medical researchers to produce a code because of their exposure to lawsuits, also argued that regulations could become a weapon wielded against physicians. There was, he warned, a "disturbing and widespread myth" that codes would provide some kind of security when, in practice, "the prosecution could show failure to comply fully, and then an endless vista of legal actions opens up."[47]

Although the medical community contested the value of formal codes, it was unable to produce a coherent alternative system that would allay the concerns of the public. It maintained differing systems of ethical practices

and beliefs that followed a fault line with clinicians on one side and researchers on the other, although some proponents of research like Beecher
straddled both sides of the divide. Yet, for all researchers' and clinicians'
differences, the answer always resided in the proper behavior of individual investigators who retained their medical prerogatives. For medical
researchers, their experimental practices were driven by a utilitarian ethics
wedded to the principle of equipoise. Research projects were proper and
ethical if there was a possibility of a net gain for society. Physicians participating in a clinical trial were ethically justified in enrolling patients if
they had no rational reason to favor any of the treatment options.[48] The
ethical principles followed by physicians who were not participating in a
clinical trial (and, for that matter, the many researchers who operated in a
clinical, as opposed to an investigational, mode) were quite different. Their
ethics were supported by the Hippocratic principle of doing no harm, a deontological imperative that was realized fiducially through the physician's
relationship to the patient. To many clinicians, this stance appeared to be
in direct conflict with the utilitarian ethics of researchers.

The two sides were in fundamental disagreement on these ethical issues, and that divide was widened by the professional and epistemological
differences between them. Physicians were quite aware during the period
following the war that medical research was growing rapidly in the laboratory and in the clinic with large-scale trials. Clinicians were conscious
and wary of a new breed of medical researcher who was beginning to dominate the field. Beecher's writings from the late 1950s on pay heed to the
rapid growth of medical research in size and in power within the medical
community: "of transcendent importance" to this development was "the
enormous increase in available funds" for research. This, Beecher argued,
had led to a world in which "medical schools and university hospitals are
increasingly dominated by investigators." And every young medical doctor
knew that he would "never be promoted to a tenure spot, to a professorship
in a medical school, unless he has proved himself as an investigator."[49]

Clinicians understood and resisted what they perceived to be the growing dominance of medical research and the type of ethics that followed in
its wake. These concerns extended beyond the U.S. community. "We have
asked ourselves," the secretary of the WMA summed up in a 1960 debate,
"whether we should continue to be guided by the Hippocratic tradition
or whether we are going to write ourselves a new code of ethics in which
experimental procedures are permitted that would have horrified our fa-
thers?" It was demoralizing, he quoted a physician, "when clinical practice
is made secondary to research" and when researchers spend "more time

in the laboratories than at the bedside." The contrast for such physicians was between medical research that could "lead to a disregard of the moral values which many consider to be absolute and not relative" and clinical practice, where medicine "always has been a dialogue between physician and patient."[50]

None of the ethical positions espoused by the medical community through the mid-1960s (whether utilitarian or deontological) would effect a change in ethical practices. Clinicians and many researchers as well as medical institutions continued to argue for a Hippocratic ethics of doing no harm that was practiced within a patient-physician relationship. At the same time, clinical trial research continued unimpeded by Hippocratic considerations and followed a utilitarian ethic in spite of the fact that many, if not most, clinical trial physicians at times espoused the Hippocratic tradition. The two traditions (utilitarianism and Hippocratic ethics) were, however, precariously held together by equipoise, which provided a rationale for clinicians to support utilitarian goals if they believed that their patients had an equal chance with any of the options of the study.

Henry Beecher suggested another approach, one derived from neither the utilitarian nor the deontological tradition. Beecher, who was deeply committed to research and troubled by its excesses, sought an answer in a virtue-based ethics that would allow research to flourish free from rigid regulations at the same time as it allowed therapy to prosper because of the virtuous behavior of investigators. For Beecher, ethical transgressions in research (as well as clinical misconduct) arose, not from bad faith, but from ignorance: "The breaches of conduct which have come to my attention were owing to ignorance or thoughtlessness. They were not wilful or unscrupulous in origin." Education was an answer, and Beecher's own writings would contribute to them since they could "help investigators protect themselves from the errors of inexperience."[51]

Henry Beecher's approach through the education of virtuous investigators would, even if it had been adopted, have done little to change practices since, if anything, his ideas required training over a long period of time and they lacked a specific procedural formulation. Moreover, all the espoused solutions were ones in which the prerogatives of physicians were left unimpaired. Beecher's virtue-based ethic resided within individual investigators, as did the Hippocratic patient-physician relationship. Even the ethics underwriting clinical trials were based on the assumption that investigators properly designed their studies and on the ability of participating physicians to ethically judge whether their patients should be entered into the trials. None in the medical community understood how to design a regulatory system that would constrain physicians' ethical behavior in a manner similar to the way in which their scientific procedures were governed within clinical trials.

Governing Medical Research and Restoring Trust

By the late 1950s and early 1960s, public concerns with medical research were increasing. The medical community's attempts at regulating research were limited and fragmented, and little had been done at the institutional level. An NIH-funded survey in the early 1960s determined that fewer than 20 percent of medical research institutions had any guidelines whatsoever for research and that almost none had any for clinical research, confirming the findings of a somewhat earlier survey of Welt's.[52] At the international level, the WMA began deliberations on a code of ethics in the mid-1950s, but did not publish a draft code until 1961, and then took another three years to put forward its 1964 Helsinki Declaration.[53] Although the code was finally able to define nontherapeutic research and distinguish such research (on healthy volunteers) from clinical research on patients, it could not provide a blueprint for regulating research practices. For example, on the important issue of informed consent, it rejected the prescriptive declaration of the Nuremberg Code demanding consent under all conditions, which it saw as hindering the prerogatives of physicians. Rather, the Helsinki Declaration left it to the judgment of physicians to obtain consent "if at all possible" and when, in the physician's judgment, it would not affect the therapeutic situation.[54] What Helsinki did was to encode extant practices. But the practices that it encoded, which left ethical judgments primarily in the hands of physicians, were inadequate to control research enough to recover and maintain the eroding public trust in medical research. The movement toward some consensus on how to govern research was so limited that the noted American legal scholar William Curran expected that changes would occur only through a gradual development of case law to establish research guidelines, rather than through the medical establishment.[55] None who were concerned with clinical ethics, Curran later noted, discussed or expected "the possibility of regulation of medical research by the federal government."[56]

Nonetheless, by the early 1960s, the director of the NIH, James Shannon, was concerned about the rapid depletion of public confidence in medical research, and he perceived the need to introduce some sort of regulatory structure. Shannon and other reformers in the Public Health Service (PHS) feared that, if the public's trust could not be restored, the medical enterprise comprising academe, industry, and government agencies might find it difficult to continue to flourish on the massive scale that had characterized the postwar period. But Shannon did not seek to wrest control of research from the medical community; instead, he created a system for governing NIH-funded research that left regulation in the hands of local committees following broad guidelines set by the NIH. These local institutional review boards (later called IRBs) would also serve as liaisons between government and academic medicine. They were the first example of a number of government-created oversight bodies that would control (and, in some later manifestations, investigate) the functioning of the scientific community.[57] The IRBs would be housed in medical centers under the authority of local practitioners, yet they would draw their patronage from the NIH. The review boards were mandated to assure that research proposals submitted to the NIH met its ethical rules concerning informed consent and risk-benefit analysis. Consequently, the medical community in general and the NIH in particular would gain from the mediating role of review boards. The IRBs would shield scientific investigators from direct government intervention at the same time as they filtered out proposals that might later be tarnished by public scandal. Consequently, medical research would flourish, and the NIH could assure its benefactor (the U.S. Congress) that it was funding knowledge that was beneficial and untainted.

The first move toward such a system was set in motion by contingent events. During congressional hearings in the early 1960s under Senator Estes Kefauver intended to investigate and control drug advertising, the thalidomide scandal broke. There were sensational revelations about mutilating defects in children born to mothers who had used thalidomide to control morning sickness, and they brought an intense public outcry. The thalidomide scandal galvanized and redirected the Kefauver hearings and directly led to the Kefauver-Harris Bill of 1962, which mandated the FDA to regulate the drug industry more thoroughly and, in particular, to require drug companies to provide proof of drug efficacy as well as safety. This meant that extensive clinical trials were required, and, in addition, the bill contained a comprehensive code regulating the clinical testing of new drugs to monitor industry compliance. Although the FDA attempted to introduce strict controls, especially on consent, by 1967 its efforts were watered down through lobbying by the pharmaceutical industry and the medical community. The regulations were brought more into line with the needs of medical researchers; the preamble to the FDA's 1967 rules contained a statement that the regulations were consistent with the Ethical Guidelines of the AMA and the Helsinki Declaration.

The situation at the NIH was quite different. Unlike the FDA, the NIH was, not a regulatory body overseeing practices, but a government body with close ties with academic medical researchers. As an arm of the PHS, it primarily funded extramural research through peer review, a system that depended heavily on outstanding medical researchers from the academic community donating their time to ad hoc study sections. It shared with medical investigators a laissez-faire ethos of medical research that relied on objective and unbiased reviews of research proposals by scientific peers. Indeed, its peer review system was the envy of scientists throughout the Western liberal democracies for its ability to fund research without overt conflicts of interest.

James Shannon, who directed the NIH from 1955 to 1968, was concerned about the ethical conduct of research that had predated the FDA regulations. As early as 1960, he had funded a three-year study of actual clinical research practices.[58] Moreover, he became particularly distressed by the notoriety that some PHS- and NIH-funded experiments had received in highly publicized press accounts of unethical practices.[59] In particular, Shannon was concerned that such experiments were carried out without a review by peers and based only on the judgment of the medical investigator. Nevertheless, he was initially reluctant to intrude into the patient-physician relationship, a position that was shared by most of his staff at the NIH.[60]

By late 1963, following the Kefauver-Harris bill, Shannon set up an in-house group (the so-called Livingston Committee) to consider what role the NIH should play in regulating research.[61] In a February 1964 memorandum, the committee stated: "The moral and ethical aspects of clinical investigations lie primarily within the relationship of the responsible physicians and associated investigators and the patient or subject to be studied."[62] In his response, Shannon did not dissuade the committee from this position.[63] The committee's first report, which followed, noted the growing number of researchers and the increasing scope of medical experimentation and warned of "the possible repercussions of untoward events which are increasingly likely to occur." It further argued: "The reputation and public confidence respecting individuals, institutions, and the NIH alike could well be rudely shaken by events that are impossible to prevent and by unseemly practices impossible to control." In spite of these concerns, the committee was unwilling to call for NIH involvement in ethical oversight since that would likely "inhibit, delay, or distort clinical research." Under pressure from Shannon, it held additional meetings and finally encouraged the NIH to formulate its own principles of clinical ethics. By that point, Shannon and others at the NIH no longer believed that the judgment of the investigator was sufficient for reaching ethical conclusions about human investigations. Shannon used the cautious recommendation of his internal committee, which stated that human subjects research should be "identified for special consideration,"[64] to press the NIH's advisory health council for "an institutional framework in which research plans and intentions involving clinical investigations are submitted for review by the investigator's peers."[65] The council's December 1965 policy statement made it clear that the government would provide research funds only if "the judgment of the investigator is subject to prior review by his institutional associates."[66] The surgeon general's order that followed on February 26, 1966, defined the scope of peer review: "The review should assure an independent determination: (1) of the rights and welfare of the individual or individuals involved, (2) of the appropriateness of the methods used to secure informed consent, and (3) of the risks and potential medical benefits of the investigations."[67]

This order had quite substantial implications since it produced a system of governance under which trust in the ethical character of research would reside in the ability of medical/government institutions to control medical practitioners rather than the ethical probity of individual investigators. Yet the move seemed almost imperceptible at the time, and, although there were complaints from the behavioral research community, the majority of medical investigators accepted the system. Indeed, the influential *New England Journal of Medicine* congratulated the PHS for its restraint.[68] Yet the implications of the surgeon general's order were significant and far-reaching. To begin with, almost all medical research came under the new system since any institution that received NIH funds was literally forced to place all its research under peer review in order to avoid a dual system of ethical practices. Moreover, the structure of the NIH's peer review system had implications that extended beyond what Shannon had, apparently, intended. Although the local review committees were initially asked to evaluate the *judgment* of the investigator, the requirement that they assess risks and benefits meant that the proposal's scientific merits and likelihood of success had to be considered. This type of review led to a subtle but important shift from judging the ethics of the medical investigator to judging the merits of the proposal. In the process, the investigator was decentered in favor of the peer review committee. As the new system of review evolved, investigators also found it convenient to transfer their ethical responsibilities to experts on the local review committees, who would produce a judgment of the adequacy of the consent statement and the balance of risks and benefits of the proposal. The troubling issues that investigators faced when experimenting on humans could be partially alleviated once the review committee had sanctioned a research proposal. The move from individual to institution meant, not only that trust been transferred away from the investigator, but also that, in an important sense, so had responsibility.

Even the responsibility of the patients had been somewhat displaced by the new system. Unlike the FDA structure, where the assessment of risks and benefits was placed within the consent requirement itself, thereby leaving judgment primarily in the hands of the patient, the NIH structure removed risk-benefit from consent and placed it under the purview of the local committee. Consequently, the welfare of the patient became a greater responsibility for the review committee. Since these risk-benefit and consent issues had to be continually assessed by peer review committees, specialized experts (medical ethicists, bioethicists) gained new roles in ethics.

The NIH system of governance also had a much more important effect on research in the United States than did that of the FDA. Part of the reason was simply numbers: in 1967, the NIH received something like 21,000 research applications, while the FDA received only 671 new drug applications.[69] Another aspect of the extent of the NIH's influence was, as we have seen, that most institutions were obliged to judge all their research proposals by the NIH standards. In addition, the regulations rapidly influenced research practices. In 1966, at the start of the new regulations, the NIH determined that 7.4 percent of the applications involved potential hazards to the patient. Within less than two years, that figure had dropped to just 1.7 percent.[70] The program affected research practices so quickly because of the broad and flexible system of regulation that the NIH had put in place. A research proposal sent to it merely had to contain a confirmation that a local peer review committee had approved it. Moreover, the review process itself was intended to leave sufficient latitude to the wide range of state and local laws and customs. For example, the form and content of the consent statement was entirely left to the judgment of the local review committees. To be sure, during the 1970s, the NIH began to specify the contents of consent statements, but even those changes left sufficient latitude in order not to impede research practices.

Flexible standards had, as we have seen, contributed to the successful control of multicenter trials where it was possible to carry out therapy and diagnosis over a wide range of institutions that made room for local needs without compromising the protocol of the trial. In a similar way, the NIH peer review system was also successful in controlling practices over a wide area because it left sufficient decision making in the hands of the local medical community while retaining the monitoring of local ethical oversight. In creating such a metrological system, the NIH brought a new standard to clinical ethics practice: the conduct of human experiments required the sanction of an investigator's peers.

The standards and practices of ethical review in the 1960s should be kept in mind when turning to Saenger's studies in the later chapters. Some critics have attacked Saenger for not obtaining consent (at least prior to the mid-1960s), while others have criticized him for the inadequacy of the disclosure of the purposes and risks of the research. But we should remember that the NIH program did not begin until 1966 and that various surveys at the time clearly showed that few institutions had any system of ethical review. *Consent* also had a very different meaning before the bioethical revolution of the 1970s. In the case of the NIH, consent had a well-circumscribed purpose: it was meant as much to provide a mechanism for patients to evaluate their risks as it was to encourage researchers to moderate their behavior. The Livingston Committee's report was quite specific on this matter. It was generally recognized, it argued, that "the physician investigator reduces the risk to his patient, to himself, and to his institution if he shares responsibility for what is done and what is not done with other competent professionals."[71]

Following the bioethical revolution, *consent* took on a much different and more universal tenor. It came to mean primarily a way to assure the protection and affirmation of the autonomy of patients. In their highly influential *History and Theory of Informed Consent*, Faden and Beauchamp envision the beginning of this later view in the NIH regulatory framework, arguing that both were "buds on the same stem." To Faden and Beauchamp, it is "indefensible" to take the position that "informed consent was exclusively directed at protection of subjects from risk rather than at protection of autonomy." As I have argued here, the NIH program aimed to reduce risks to the medical community, the NIH and patients. Consent was instrumental to those goals. Indeed, throughout the period during which the NIH developed its scheme, discussion of consent at the NIH remained almost entirely in the background.[72] For researchers like Saenger who were educated following the war *consent* could mean only, as it did for their patients, a means for alerting subjects to the risks and benefits. Most important, a paternalistic patient-physician relationship governed disclosures in the 1960s so that it was not uncommon for physicians to withhold from patients the details of their therapy and, even in some cases, certain truths about their illnesses.[73]

The mid-1960s provides an important, even radical, break in the governance of clinical conduct. The research practices of physicians would now require formal review by peers, who would judge the risks and benefits of the research and the form and content of the consent statement. Local IRBs would begin to play an important role in the production of medical knowledge. Physicians would have to receive the approval of these boards in order to carry out any significant research on human subjects. Although, at the start, these reviews were in many cases somewhat perfunctory and limited in scope, over time the structure and content of the review process became quite detailed and bureaucratic. At the University of Cincinnati, local peer review was formally implemented following the 1966 PHS announcement regarding ethical regulation. Saenger's total-body irradiation experiments came under the purview of a local peer review committee, and chapter 6 follows, in some detail, the committee's attempts to tame his program. Before discussing Saenger's program, however, I need first to look at the relationship between cancer therapy and military medicine that developed in a number of places in the postwar period. Thus, in the next chapter, I follow one major investigational effort, the development of a treatment for childhood leukemia, and show how, even in this specialized area, military concerns played an important role.

Notes

1. Nuremberg Code, *Trials of War Criminals*, 181–82.

2. Shuster, "Fifty Years Later," 1436.

3. Shevell, "Neurology's Witness," 277.

4. See, e.g., Perley et al., "Nuremberg Code"; and Winslade and Kraus, "Nuremberg Code Turns Fifty."

5. See, e.g., Katz, "Consent Principle," 227–39.

6. Rothman, *Strangers at the Bedside*, 78–92; Winslade and Krause, "Nuremberg Code Turns Fifty," 148; and Jasanoff, *Designs on Nature*, 175. The Beecher article is "Ethics and Clinical Research."

7. Rothman, *Strangers at the Bedside*, 90.

8. Shapin, *Social History of Truth*, 410–13.

9. For excellent reviews, see, e.g., Curran, "Current Legal Issues"; Faden and Beauchamp, *Informed Consent*; and ACHRE, *Final Report*.

10. Beecher, "Experimentation in Man," 109.

11. For human experimentation and ethics prior to World War II, see Lederer, *Subjected to Science*. For a general history of medical ethics and human experimentation, see Howard-Jones, "Human Experimentation"; and Jonsen, *Short History*.

12. ACHRE, *Final Report*, chap. 2.

13. Ibid.

14. Quoted in Beecher, *Research and the Individual*, 221.

15. ACHRE, *Final Report*, 77.

16. Ibid., 78.

17. Shuster, "Fifty Years Later," 1439. For further claims that Ivy "flirted...with perjury" on the witness stand, see Harkness, "Significance of the Nuremberg Code."

18. For a discussion of the yellow fever experiments and the use of consent, see Lederer, *Subjected to Science*, 131–35.

19. Shimkin, "Problem of Experimentation," 205.

20. Guttentag, "Problem of Experimentation," 209.

21. Wiggers, "Human Experimentation," 122.

22. Shimkin, "Problem of Experimentation," 207.

23. Beecher, "Experimentation in Man," 461.

24. Legal Environment, "Report on the National Conference," 139.

25. Ibid., 140.

26. Welt, "Reflections," 130.

27. Beecher, "Human Experimentation," 79.

28. Quoted in Curran, "Governmental Regulation," 403. During the period under discussion here, physicians would have made reference to either the 1949 second edition or the 1956 third edition of Regan's *Doctor and Patient and the Law*.

29. Faden and Beauchamp, *Informed Consent*, 191.

30. Curran, "Current Legal Issues," 297.

31. Faden and Beauchamp, *Informed Consent*, 129–32. Another much-quoted case is the 1957 *Slago v. Leland Stanford Jr. University Board of Trustees*. Martin Slago suffered permanent paralysis as the result of a translumbar aortography. He successfully sued for negligence (using the *Schloendorff* case of 1914 and others that applied to battery) to argue that the physicians had a duty to disclose "any facts which are necessary to form the basis of an intelligent consent by the patient to the proposed treatment" (quoted in Faden and Beauchamp, *Informed Consent*, 125).

32. Curran, "Current Legal Issues," 297. According to Jasanoff (*Science at the Bar*, 33), in 1960 the courts considered it an affirmative duty for physicians to disclose risks, but it was not clear which risks should be disclosed.

33. Beecher, "Human Experimentation," 79, 80.

34. Shimkin, "Problem of Experimentation," 207.

35. Legal Environment, "Report on the National Conference," 138.

36. According to Robert Baker, the 1847 AMA Code was a combination of John Gregory's emphatic medical humanism and Thomas Percival's conception of the physician's office as a public trust, with the addition of the duty of physicians to serve society, even at their own jeopardy. The AMA simplified Percival's complex hierarchical ethics into a tripartite contract between physicians and patients, physicians and other physicians, and physicians and society. Physicians should minister to the sick, and, in exchange, patients should follow their physician's prescriptions. Physicians should not speak out against one another for the welfare of the profession. Finally, society should respect, i.e., license, physicians as recompense for their duty and obligations to the welfare of society. The code states that it is physicians' duty to be vigilant regarding the welfare of the community and, when pestilence prevails, "to face the danger and to continue their labors for the alleviation of suffering, even at jeopardy to their own lives" ("History of Medical Ethics," 869).

37. For an argument of medicine's insularity in the postwar period, see Brandt and Freidenfelds, "Research Ethics after World War II," 240–43.

38. The situation in Block 50 appears to have been different from the way it was portrayed in Ivy's antitotalitarian diatribe. Ludwik Fleck, the famous microbiologist and historian and philosopher of medicine, had been interned in December 1943 in Block 50 of Buchenwald, part of the SS Hygiene Institute, to develop a serum against typhus. On arriving, Fleck discovered that the prisoners had isolated, not rickettsia, but another type of germ. According to one of the prisoners: "We asked him [Fleck] not to convey this to Ding [the SS head of Blocks 46 and 50] but to experiment with us and help get us out of this difficulty. During the two years that he worked with us Dr. Fleck kept the secret." This secret was maintained by supplying the two types of vaccine. After the war, in 1946, Fleck published an article, "The Problem of the Science of Science," in which he made an epistemological analysis of the scientific work in Block 50. He argued that a collective truth had been reached that was consistent, though wrong owing to a systematic error that resided outside the scientific collective of the prisoners. Fleck also appeared as a witness at Nuremberg against the IG Farben Co., which had been involved in experiments in the notorious Block 46, where typhus had been injected into prisoners (Schnelle, "Microbiology and Philosophy of Science").

39. Ivy, "Nazi War Crimes," 268.

40. Ibid., 271.

41. Shimkin, "Problem of Experimentation," 205.

42. Beecher, "Experimentation in Man," 109.

43. Legal Environment, "Report on the National Conference," 138.

44. Both quoted in Welt, "Reflections," 128.

45. Ibid., 129.

46. Beecher, "Experimentation in Man," 120.

47. Beecher, "Consent in Clinical Experimentation," 34.

48. Freedman, "Equipoise," 141. See also Fried, *Medical Experimentation*, chap. 1.

49. Beecher, "Ethics and Clinical Research," 1355.

50. Secretary WMA, "Editorial," 95.

51. Beecher, "Experimentation in Man," 109. Beecher was taking a Platonic position: if individuals know the good, they will always choose it. Virtue-based and Hippocratic deontological ethics existed hand in hand among physicians. For their long and complex history in American medicine, see Baker, "History of Medical Ethics."

52. Curran, "Governmental Regulation," 406–9. The survey was carried out by the NIH-funded Law-Medicine Research Institute at Boston University following the 1958 National Conference on the Legal Environment (Welt, "Reflections").

53. Reprinted in Beecher, *Research and the Individual*, 277–79.

54. Ibid., 278.

55. Curran was at the time the head of the Law-Medicine Research Institute.

56. Curran, "Governmental Regulation," 409.

57. I have been influenced here by the work of Guston (*Between Politics and Science*, 96–98), who presents a nuanced argument for a contractual relationship between the scientific community and the government. Guston argues that the contract broke

down in the late 1970s when direct government control was imposed through new agencies housed *outside* the scientific community. These agencies would act as oversight bodies but maintain distinct boundaries between the government and researchers. Here, I am arguing that, while IRBs operated *within* the medical community, they were, nevertheless, serving similar oversight and boundary-keeping roles by the middle of the 1960s. For a discussion of the role of local review in Cincinnati, see chapter 6.

58. Frankel, "Development of Policy Guidelines," 48. The NIH funded the Boston University Law-Medicine Research Institute, which had broad-ranging interests in medical research ethics.

59. For example, the implantation of an animal kidney into a human and the injection of live cancer cells into indigent patients (ACHRE, *Final Report*, 99).

60. Frankel, "Development of Policy Guidelines," 47.

61. The committee was named after Robert Livingston, the associate chief of program development.

62. Robert Livingston to James Shannon, February 20, 1964, "Moral and Ethical Aspects of Clinical Investigation," http://search.dis.anl.gov/ (accessed July 2001; hard copy in author's files).

63. James Shannon to Robert Livingston, March 5, 1964, "Review of the Ethical Aspects of Clinical Investigation," http://search.dis.anl.gov/ (accessed July 2001; hard copy in author's files).

64. Robert Livingston to James Shannon, November 4, 1964, "Progress Report on Survey of Moral and Ethical Aspects of Clinical Investigation," 3, 7, 9 (on encouraging the NIH to formulate its own principles), 10, http://search.dis.anl.gov/ (accessed July 2001; hard copy in author's files).

65. James Shannon to Surgeon General, January 7, 1965, "Moral and Ethical Aspects of Clinical Investigation," 2, http://search.dis.anl.gov/ (accessed July 2001; hard copy in author's files). There has been much made about the influence of Henry Beecher and how his exposé of unethical research led to the new NIH and FDA regulations. Winslade and Krause ("Nuremberg Code Turns Fifty," 148) argue that Beecher's 1966 "Ethics and Clinical Research" "ultimately lead [*sic*] to unprecedented controls on research." Sheila Jasanoff makes the stronger claim that the NIH and the FDA responded almost at once to Beecher's article (*Designs on Nature*, 175). Rothman (*Strangers at the Bedside*, 70–84) also places great weight on Beecher's work.

66. Frankel, "Development of Policy Guidelines," 52.

67. USPHS to Clinical Investigators Using Human Subjects, February 8, 1966, http://search.dis.anl.gov/ (accessed July 2001; hard copy in author's files).

68. Frankel, "Development of Policy Guidelines," 52.

69. Curran, "Governmental Regulation," 432.

70. Ibid., 439.

71. Robert Livingston to James Shannon, November 4, 1964, "Progress Report on Survey of Moral and Ethical Aspects of Clinical Investigation," 5, http://search.dis.anl.gov/ (accessed July 2001; hard copy in author's files).

72. Faden and Beauchamp, *Informed Consent*, 207.

73. During my tenure at Thomas Jefferson University Hospital, I would spend one day per week at a hospital in Bryn Mawr, an affluent suburban community outside Philadelphia. At the time—the early 1980s—it was still considered improper (bad manners) to mention the term *cancer* in front of patients.

3 Military Medicine and Cancer Therapy

It is now difficult to imagine what the social institution of science would look like divorced from its military ties. • Steven Shapin, "Science and the Public"

At a 1994 House of Representatives subcommittee hearing regarding cold war radiation experiments, Eugene Saenger defended his work with total-body irradiation (TBI) by aligning himself with a long history of traditional medicine. He claimed that his program was therapeutic and used TBI and bone marrow transplants to treat advanced cancers. As part of his defense, he linked his investigations to those of other researchers who had applied similar combinations of radiation and bone marrow transplants to treat lymphoma, leukemia, and other disseminated diseases. He also aligned himself with studies both in Cincinnati and elsewhere that followed his program and used TBI to treat patients with painful disseminated disease.[1] Saenger saw his program as deeply enmeshed in a network of studies in which large-field irradiation was applied to difficult cases in cancer therapy.[2]

Shortly following these hearings, President Clinton's Advisory Committee on Human Radiation Experiments (ACHRE) positioned Saenger in an entirely different historical setting. To ACHRE, he was not part of the cancer research community but, rather, one of a number of investigators who were funded by the Department of Defense (DOD) to answer purely military questions: for example, what the physical and psychological effects of TBI would be on military and civilian populations in the event of a nuclear attack. As a consequence (or so the argument runs), these investigators—and Saenger was the most notorious of them—exploited their patients since the use of TBI therapy held few, if any, benefits for them.

These historical reconstructions are grounded on two important considerations regarding the relationship between cancer research and military medical research. First, there is an underlying assumption of essentially two distinct and separate communities: one populated by investigators working on purely cancer research questions, the other populated by investigators who sought to answer primarily military questions. Second, the boundary separating these domains permitted information to flow in one direction only. That is, military investigators learned about and took on board some of the ideas and practices that had been developed for cancer therapy, while cancer researchers had little or no interest in developments in military medicine. Ideas and practices flowed from the cancer therapy to the military medicine domain, not in the reverse direction.

As we will see, these assumptions do not hold. Cancer therapy research drew significantly on advances in military medicine, and, most important, the boundary between the two disciplines—cancer therapy and military medicine—was quite diffuse and interaction across it dynamic. In particular, military studies concerned with diagnosing and treating radiation injury had a profound influence on the development of techniques for treating acute leukemia with TBI and bone marrow transplants. Standard medical accounts of the development of this latter treatment do not acknowledge such influence. The histories usually suggest that, in the years following the discovery of X-rays until the Second World War, TBI with low-energy X-ray sources yielded little benefit for the treatment of leukemia. Following the war, the introduction of high-energy radiation sources like cobalt 60, more potent chemotherapy drugs, and new technologies for matching and handling bone marrow provided a new potential to treat leukemia. These standard narratives then invariably jump to a groundbreaking 1977 report on the first hundred patients treated for leukemia with bone marrow transplants in which Donnal Thomas claimed to be able to obtain long-term remissions.[3] The crucial role played by military research in the diagnosis and treatment of radiation injury as applied by cancer therapy investigators is invariably left out of such histories.

If we take account of this mutual and interdependent history, we begin to see a much more complex relationship between cancer and military-related research during the postwar period. Investigators, whatever their primary research goals, shared ideas, technologies, and practices; they also inevitably followed similar protocols for ethical conduct. Indeed, since a number of researchers at different times investigated both cancer-related and military-related problems, they would have been likely, as I discuss in the final section of this chapter, to follow similar standards of clinical conduct.

In the first section of this chapter, I review how the atomic bomb attacks on Hiroshima and Nagasaki led to the framing of a new medical entity, the bone marrow syndrome, and how it provided a focus for a wide range of laboratory and human studies in both cancer therapy and military medicine. In the second section, I look at research in the "traditional" cancer therapy community. In particular, I follow the work of Donnal Thomas and his use of TBI and bone marrow transplants to treat acute leukemia. This section makes it clear that ideas and techniques related to TBI circulated freely among military and cancer investigators, whatever their primary research aims. The work of Thomas also provides us with a case history of medical research in the postwar period to compare to the work of Bernard Fisher and Eugene Saenger. Here, as in the other two cases, we find that the process of research was messy and that investigative directions had to be continually adjusted to account for contingent events. Thomas's research was highly dynamic, and, at the same time, he followed the kind of clinical trial ethics of his peers that was discussed in the previous chapter.

In the last section, I consider radiation studies devoted to military-related questions that were carried out primarily by cancer therapy researchers. These human radiation studies began in the early postwar period and were funded by the DOD. By the early 1960s, a number of medical facilities, including the one in Cincinnati that housed Saenger's program, used cancer patients as proxies for soldiers to answer military questions. In this section, I also argue that it is not straightforward to claim that the research ethics of these military funded studies were fundamentally different from the ethics that informed traditional cancer research.

The rigid borders that presumably existed between investigations carried out for military and those carried out for cancer therapy goals, including the ethical norms that guided the researchers, have been the product of later boundary work. Such efforts have distorted the intimate relationship between military and cancer research during the cold war, and Saenger's studies should be viewed in light of this more complex world of medical research.

Radiation Sickness and the Bone Marrow Syndrome

In the period immediately following the atomic bomb attacks on Hiroshima and Nagasaki, a new type of illness was defined. Radiation sickness, as it was called, was studied by the Atomic Bomb Casualty Commission (ABCC), which began its work within weeks of the atomic blasts. Radiation sickness, a previously unknown constellation of illnesses, was characterized by three different syndromes, depending on the distance of the victims from ground zero. Prior to atomic warfare, physicians who treated leukemia and other disseminated diseases with TBI recognized that it could have appreciable effects on various tissues, including the blood-forming systems. That work had extended as far back as 1907, when Dessauer proposed applying a radiation bath to the entire body of a patient.[4] Although TBI treatments extended up to the Second World War, so little was understood about radiation effects that human and laboratory investigations were initiated out of concern for workers on the Manhattan Project.[5] Even so, the circumstances and context of the radiation studies prior to and during the war were so narrow and limited that there was scant understanding of the clinical consequences of the high doses that were experienced by the Japanese victims. James Oughterson and Shields Warren, the editors of the ABCC report on the medical effects of the atomic bomb, remarked: "Until the atomic bombs exploded in Japan, there was almost no information on the lethal and sublethal effects of total-body exposure. Certain general conclusions had been drawn from experimentation with animals, but the details of the clinical syndrome were as yet unknown."[6] Oughterson and Warren refer to a situation that was unprecedented. Not only were there many immediate victims of the blasts, but others, who initially survived and appeared to have no complications, also later experienced appalling, and often fatal, symptoms. In her book on the ABCC, Susan Lindee describes it thus: "Many survivors who came through the blast unhurt began, within minutes, hours or days to manifest the acute symptoms of radiation sickness. Sudden severe nausea and diarrhea were the first signs, followed later by subcutaneous bleeding and gingivitis. Many of the seemingly unhurt survivors died within days and weeks after the bombing of an illness Australian Wilfred Burchett called 'atomic plague.'" No one understood what was happening, neither the Japanese physicians nor the American military. The Manhattan Project scientists (those involved in developing the atomic bomb under the U.S. Army Corps of Engineers) initially dismissed Burchett's claims as merely Japanese propaganda. The response, however, was not entirely political. Lindee claims that American authorities were genuinely skeptical since Manhattan Project scientists at Los Alamos felt that Burchett's story "could not possibly be correct."[7] No one, even those most intimately involved in the development of the atomic bomb, had much understanding of the effects of radiation on humans beyond the immediate acute response.

In the months and years following the attacks, every conceivable aspect of the effects of the atomic explosions were investigated by American and Japanese scientists. The dosimetry of the explosion was laboriously reconstructed on the basis of the radiation characteristics of the bombs, the wind conditions, the location and movement of the victims, the presence or absence of shielding, and so on. The victims' clinical symptoms were monitored, categorized, and correlated with the calculated dose patterns. In spite of the enormous difficulties and the evident uncertainties, researchers developed a multitier topography of radiation sickness in which each region was demarcated by dose. Starting at the upper end, in the vicinity of one thousand rads[8] or more, victims began to suffer, within minutes or days, from convulsions and coma, and death inevitably followed from what was termed the *central nervous system syndrome*. The *intestinal syndrome* would begin around half that dose, and, here, the victims would have progressive nausea and vomiting that, at higher doses, would increasingly lead to infection and death.

Those who did not initially succumb to these syndromes and had received lower doses might have experienced few, if any, effects, but, within weeks, they began to suffer from fevers, bleeding, mouth ulcers, and hair loss. Warren noted that many of the victims "had massive hemorrhages from various body orifices." The ABCC investigators soon concluded that the bleeding was not due to the destruction of blood platelets directly by radiation since, if that were so, Burchett's atomic sickness would have manifested itself immediately following the blast. Bleeding was, rather, a delayed consequence of the failure of radiation-damaged bone marrow to replenish the natural depletion of circulating platelets: "It is fair to assume that the absence of hemorrhage deaths in the early days suggests that the blood platelets in the circulation were not destroyed by radiation and only as a low point was reached as a result of deficiency in blood did hemorrhage manifestations occur." Warren suggested that the best treatment for bone marrow sickness was supportive therapy, although his suggestion was put in negative and contemptuous terms: "The treatment given by the Japanese was utterly inadequate; . . . repeated blood transfusions, penicillin to control infection during the leukopenic period should have materially reduced the number of deaths."[9] Warren's coldness and disdain is, in many ways, typical of a number of American scientists in Japan in the early postwar years.

Following the attacks on Hiroshima and Nagasaki and especially after the Soviet Union tested a nuclear device in 1949, government support for radiation studies rose rapidly as a consequence of fears of nuclear attack on civilian and military populations.[10] The funding for the research came predominantly from the DOD, the Atomic Energy Commission (AEC), and, to a lesser extent, the National Cancer Institute (NCI). In theory, DOD funding was meant to support short-term research that would answer immediate questions that were of importance to the armed services, while AEC funding was intended to apply to longer-term, more fundamental research.[11] But, in practice, the funding situation was more complex. In part, this was due to the complicated remit of the AEC. When it was created in 1946, the AEC was required to oversee the peaceful and military development of atomic energy. By the late 1940s, it also began to consider biomedical problems, although these were primarily aimed at resident workers. However, by the late 1950s, the AEC had developed, along with its military and industrial roles, a formal and extensive biomedical program.[12] The range of its concerns meant that research monies and ideas could move across its internal boundaries. But sources of funding also came from the medical community, including the NCI, which was mandated to support cancer therapy but would sometimes support research that had primarily military goals.

Radiation sickness, and, more specifically, the bone marrow syndrome, was one of the paradigmatic problems that could be used to tap into this varied mix of funding sources. Radiation sickness became the focus of laboratory researchers, resulting in the production of an enormous amount of work of military and civilian concern. Indeed, the wealth of research on radiation sickness in the early 1950s was so staggering that an authoritative 1952 review article listed 156 references on TBI for the period January 1950–September 1951 alone. The article's authors also claimed that that number barely skimmed the surface: "We have attempted to include primarily those papers which have advanced our present concepts of the clinical manifestations of radiation injury. A complete list of references would fill the entire allotted space."[13] The referenced studies—predominantly TBI experiments on animals—involved numerous medical fields, some with long prior histories like infectious diseases, cytology, immunology, endocrinology, physiology, and some of more recent vintage, including radiation biology, health physics, and industrial medicine. To term these *total-body* or even *radiation* studies does scant justice to the range of work and the subsequent implications of the research. All this research centered around three issues: how to characterize, diagnose, and treat (or prevent) radiation injury. The primary beneficiaries of such research were, presumably, soldiers and civilians who might be exposed to TBI in the event of nuclear warfare, and, later, with the growth of nuclear power, the studies were meant to help the victims of industrial accidents.

Donnal Thomas and the Treatment of Childhood Leukemia

The paradigm problem of radiation sickness drew researchers from the military and cancer therapy communities. The work of Donnal Thomas on the treatment of childhood leukemia highlights the importance of advances in the diagnosis and treatment of radiation sickness (many by workers interested in military issues) for the development of new strategies in cancer therapy. From his initial attempts at treating leukemia patients in 1957 to his 1977 paper announcing long-term remissions, the successful treatment of the bone marrow syndrome was critical to his success. Thomas's story will reveal that his early studies were not solely aimed at cancer therapy but also addressed military concerns. Within a short time, however, his work was almost entirely aimed at finding a treatment for leukemia, although he continued to participate in conferences whose programs were primarily about radiation studies for military and industrial applications. Thomas gained much information from those conferences as well as support and criticism, while his own work contributed to research in military and industrial medicine.

One of the most important areas that affected—indeed, launched— Thomas's leukemia efforts were animal experiments carried out in the early to mid-1950s by radiation biologists in medical and military-supported research centers. These studies, in which mice and rats were exposed to TBI, not only further established some of the characteristics of radiation sickness, but also suggested possible methods of preventing and treating it, methods that went well beyond supportive therapy. These investigators learned how to protect animals from an otherwise lethal dose of radiation (in the range of one thousand rads for mice) by placing a lead block over the spleen thereby reducing the amount of radiation it received. Although blocking the spleen had a limited value for the military, it led to further studies focused on attenuating or reversing the effects of TBI. Researchers injected spleen, liver, and bone marrow cells into animals *after* they received an otherwise fatal dose of TBI and discovered, surprisingly, that the injected cells had a restorative effect.[14]

What ensued was a long series of debates replete with charges and countercharges over which agent or agents were effective and why. One area of controversy was whether bleeding from radiation sickness was primarily due to a lack of platelets or, rather, a consequence of the availability of heparin-like compounds in the victims. The latter position questioned whether injected bone marrow was the active agent protecting victims from radiation sickness. A series of experiments by Eugene Cronkite and others finally led to a consensus that bleeding was controlled by the presence of sufficient quantities of platelets, not heparin-like compounds. But, to settle the matter, Cronkite had to learn how to isolate platelets and, thus, developed a platelet production unit that was later used in the treatment of leukemia.[15] Consensus was also reached that injected bone marrow was the active agent, that it had the capacity to locate itself within the destroyed marrow tissue and produce platelets and other components of the blood.

Once consensus was reached, the possibility of using bone marrow transplants to treat radiation sickness led to a series of further animal studies. Investigators demonstrated that they could obtain successful transplants by injecting marrow under a variety of conditions, from the same inbred strain, from different genotypes of the same strain, or even from different species. Animals were created that shared certain characteristics with those of the breed that had contributed their bone marrow. One researcher noted that injecting bone marrow from a rat into a mouse produced an animal that would continue to live as mouse but would have the blood tissue of a rat: "In a mouse protected against a lethal dose of irradiation with rat marrow, ultimately this mouse may have most or all of its circulating erythrocytes those of a rat, its circulating granulocytes those of a rat, its platelets those of a rat, its splenic and thymic population that of a rat, and its serum gamma globulin that of a rat."[16] Although such chimeras[17] were viable, they also began to fall prey to a new and ominous disease, the *secondary syndrome*, where the cells arising from newly grown marrow would attack their host.[18] But the disease did not appear only in animals; secondary syndrome was also diagnosed in early attempts to treat radiation sickness in humans with bone marrow transplants. For example, while carrying out research in France (supported by the AEC), Mathé reported the appearance of the secondary syndrome in bone marrow transplants of victims of nuclear accidents.[19] The syndrome would later become of critical importance when treating leukemia with bone marrow transplants, although, by that time, its name was changed to *graft versus host disease*.

It was not a far step for animal researchers at Harwell (a military-funded laboratory in the United Kingdom) to deliver, in 1956, a lethal dose of TBI with the purpose, not of simulating a radiation attack or accident, but of destroying all an animal's leukemia cells along with any normal bone marrow. After killing all blood-forming tissues, the animal, it was claimed, was reconstituted with an injection of healthy marrow from a donor, and this marrow could grow in the host free of leukemia.[20] These experiments irrevocably wedded radiation sickness to TBI and bone marrow transplants for the treatment of leukemia.

Shortly following these experiments, in 1957, Donnal Thomas and a few other researchers began to apply similar ideas and technologies in humans.[21] In his initial study, Thomas was somewhat ambivalent about the purpose of bone marrow transplants. On the one hand, he reasoned that, since radiation diseases will occur as the result of nuclear disasters, it would be important to "determine the availability and usefulness of bone marrow infusions for the treatment of these conditions in man." On the other hand, he argued that "in selected patients with disseminated neoplasia it may be advantageous to use total-body radiation in large doses and to cover the resultant aplasia by marrow transplantation."[22] Thomas was straddling the fence between applications in cancer therapy and military and industrial medicine.

Nevertheless, the attempt to treat leukemia patients with very high doses of radiation was a radical and highly dangerous move. Thomas's only guide, besides the recent animal experiments, was a trail of studies and data on radiation sickness that led back to Hiroshima and Nagasaki.[23] There certainly were no cancer therapy precedents to follow. Previous work on the treatment of leukemia and related diseases with TBI had relied on much lower doses and had more moderate aims. The early prewar role of TBI was exemplified by the noted physician James Ewing, who observed in his Caldwell Lecture of 1925 that, "in addition to inhibiting or destroying tumor tissue," TBI "may exert a favorable influence upon the nutrition and metabolism of the body as a whole."[24] Ewing's holistic arguments regarding the role of TBI persisted well into the Second World War. When the American physician Sanderson adopted the German quadrant technique, which applied TBI to one-quarter of the patient per day (presumably to be able to give higher doses), he immediately reduced the prescribed dose and claimed that constitutional improvements were "accomplished with smaller doses than are customarily employed."[25] In 1942, a team at Memorial Hospital in New York, which had the most experience with TBI, noted that, even though the survival rates of patients with leukemia were disappointing, TBI provided a "constitutional improvement" that was "manifested by a sense of well-being, betterment of gastrointestinal function, weight gain, and increase in hemoglobin."[26]

The prewar practice of using relatively modest doses of TBI had little to offer Thomas. He intended to deliver high doses, and constitutional therapy provided him with little guidance as to the onset and treatment of the potentially massive complications that would follow. His ideas shared more with the postwar heroic medical culture, which had turned, not only to the increasingly toxic chemotherapy regimens discussed in chapter 1, but also to a number of very radical surgical techniques.[27] What Thomas could, however, take over from that prewar TBI culture were certain technologies like orthovoltage X-ray machines to deliver TBI and methods for positioning the patient and delivering the treatments. Still, even here, Thomas was dissatisfied, and he would later rid himself of most of these technologies.

Thomas also drew on the limited material available on humans, mainly covering the Japanese experience and a few industrial nuclear accidents, to provide him with some (very limited) indications of the possible consequences of his intended program. By the time he began using massive doses of TBI, he was certainly aware of the dire consequences that might lie ahead for his patients, which included "the possibility of immediate or delayed reaction to the marrow, of failure of the graft, and late sequelae of heavy radiation."[28] However, he did not, and could not, know all types and the severity of complications that might result from his technique.

But, even before he got far enough to be concerned with the aftermath, Thomas faced a wide range of technical problems. He had to determine how much marrow to give, where and how to procure it, how to filter, store, and preserve it, and how to deliver it to the patients. Sometimes, the marrow would be injected following an aspiration from a living donor; in other cases, it was procured under sterile conditions post mortem and injected sometimes fresh and sometimes after freezing and thawing. Since Thomas had earlier developed techniques for handling bone marrow at Harvard, where he learned to filter it and to measure DNA activity, he was able to determine that human marrow maintained high levels of DNA synthesis up to four hours after death. He could, thus, successfully procure marrow from cadavers and filter and freeze it for later use.[29]

Thomas also had to determine whether TBI or other toxic agents were best suited for destroying the bone marrow as well as the doses he needed to give and the doses he could practically deliver without immediately killing the patients. The issue of complications from TBI was not, however, the only concern (even though it was of great importance). Thomas also worried about the complications from the transfusion itself, in addition, of course, to the most pressing concern, whether the marrow would take at all. In the latter case, death from radiation sickness was inevitable. In 1957, Thomas reported on six patients who were given marrow transplants, claiming that he might have accomplished a temporary "take" in one of them.[30] He also made a special point in the paper that the marrow could be given without embolisms in the lung, something that had clearly concerned him.

Thomas's ambitious program was filled with uncertainties—from the brief description given above we can only begin to appreciate the daunting and, at times, seemingly insurmountable technical problems that he faced. His radiation treatments were highly toxic, the transplant procedures were unstable, and he had little to guide him regarding the difficult ethical issues. His studies were of the Phase I/II clinical trial variety; that is, he was doing a safety study and efficacy study at once. This meant that he was trying to find some way of modifying the enormous number of techniques that were required so that he could reach a reproducible arrangement allowing some of the patients to survive for extended periods of time. Clinical trial practices of the period, which, as I noted in the last chapter, were becoming very aggressive, would have suggested ethical benchmarks with very wide latitude for such risky therapy on hopelessly ill patients. The only formal code of practice that Thomas could turn to—that of the American Medical Association (AMA)—was of almost no value. One of its three points was that safety should be established first in animal studies. Such a principle was clearly ludicrous when applied to Thomas's studies since most of the daunting problems he faced could in no way be tested in any system other than in humans.

Consent practices of the day, recall, were limited, and, in Thomas's case, we find no mention of them in his 1957 and 1959 papers, nor were there any as late as a 1975 paper in the *New England Journal of Medicine*.[31] It was not until his landmark 1977 paper that Thomas wrote: "All protocols were reviewed and approved by the Human Subjects Review Committee of the University of Washington and/or the Fred Hutchinson Research Center. The procedures, along with the potential risks and benefits, were explained in detail to all family members."[32] The presence/absence of informed consent seems little different in Thomas's case than in those of his contemporaries, namely, that formal consent began to be obtained and acknowledged more widely only after the NIH regulations of 1966 were put in place.

During the early postwar period, distinguishing the clinical conduct of one researcher from another on the basis of consent is clearly problematic since the absence of consent in therapeutic research was, essentially, ubiquitous. But the absence of consent documents in the historical record does not mean that the clinical ethics of researchers cannot be considered in other terms. What is most striking about Thomas is that he published many of his failures in fully documented case studies. Those studies attest to how closely he had to have experienced the life-and-death struggles of his patients. They also attest to an unusual integrity and commitment to have presented his failures in the open literature in such blatant detail since they could not add to his own standing and he could have published them only in the hope of providing valuable clues to other researchers.

Although at first Thomas had flirted with the industrial and military role of bone marrow transplants, by 1959 he had switched his attention almost entirely to leukemia. He reported on twelve cases treated with TBI and bone marrow transplantation, and, although there were examples where the marrow had taken hold and began to produce blood tissues, he could still not prolong, in any meaningful way, the lives of the patients. The problems that he and his patients faced were evident from the case reports. Patient 10 was a sixteen-year-old in relapse from leukemia who was given TBI by the quadrant technique with an orthovoltage X-ray unit operating at low intensity. She received an estimated average tissue dose of 325 rads.[33] The bone marrow transplant, which followed the radiation treatment, consisted of an injection of (3 billion) nucleated bone marrow cells procured from aspiration of the sternum and iliac crests of her sister.

During the postradiation period, the patient was constantly nauseated and vomited on several occasions. To prevent bleeding and provide gamma globulin, she received multiple transfusions using platelet-preserving equipment. She became critically ill on the twenty-first day postirradiation with a 105-degree Fahrenheit temperature, but she responded to a change of antibiotic regimen. She also developed herpes lesions, which became angry and ulcerated, but they cleared by the thirty-sixth day. A bone marrow biopsy on the same day revealed normal bone marrow, but, unfortunately, she died shortly thereafter. In other cases, Thomas reported bone marrow "takes" in which "the recipient becomes a chimera producing and tolerating cells of the blood type of the donor and in general recognizing his tissues as friendly."[34]

Whether the marrow would take, just how friendly the recipient was to the injected marrow, and how friendly the cells that carried the donor's imprint were to the host were issues that depended on numerous factors. The highly complex transplant technology consisted of multiple, intersecting processes containing many unknowns, including emerging immunological issues, as well as technical problems extending into such diverse areas as the preparation of the bone marrow and the technique by which to deliver TBI. Since Thomas had some successful takes but none of the patients survived, he believed that he was manipulating too many elements. He turned, not forward to modifying the emerging immunology, but backward to the element that had the longest pedigree, that is, TBI. This seems a curious move in retrospect since it was the immunology that was least understood and Thomas certainly acknowledged so at the time.

Yet Thomas turned toward TBI for technical and political reasons. He clearly did not like TBI at all and wished that he could be rid of it. In his 1959 paper he stated: "Total-body irradiation can scarcely be touted as an optimal treatment. Its effect on normal and neoplastic tissue is too indiscriminate to be attractive as a solution to the problem of neoplasia or to the associated problem of homograft tolerance." He was, however, stuck with TBI as an ablative agent and with a technology that he had inherited from the prewar period—and he specifically blamed the technology for the problems he was facing. Transplants, he conjectured, worked in animals but not humans because of the technical limitations of TBI delivered with orthovoltage X-ray machines: "One possible reason why success with marrow transplantation and with the treatment of leukemia has been obtained in the mouse is that the small body size and relatively delicate bony structure of the animal have permitted uniform ionization effects in the 1000r range with the use of 250 kV [orthovoltage] photons."[35] Thomas conjectured that the reason his transplants were failing was that, unlike in the situation in mice, orthovoltage radiation delivered such highly nonuniform patterns of radiation in humans that not all the leukemic cells were eradicated. In addition, the quadrant technique that he had inherited from Sanderson was inadequate because only one-quarter of the patient was treated on successive days, which might permit circulating mitotic leukemia cells to escape radiation death.

Thomas proposed to build a specialized total-body facility that would house two high-energy cobalt-60 units that would simultaneously irradiate the left and right sides of a patient. One of the units would have the rotational capability of a conventional cobalt-60 housing so that it could also be used to easily deliver radiation to any part of the body. This move was not only technical but also political. By building a specialized facility, Thomas would be able to take full control of TBI, the technology, the procedure for the delivery of treatment, and the scheduling of patients. He needed to put TBI into a "back-box" where it was well defined, predictable, and fully under his control—and he could not go on until he had done so. Thomas claimed: "Until this matter [TBI] is in hand it may be difficult to assess the immunological problems underlying success and failure of marrow transplants."[36] The contingent character of research is quite evident here. Although Thomas appeared to distrust TBI in general, the recent development of cobalt-60 radiation units and the political issues that he faced in the clinic contributed to decisions that were by no means self-evident. Given her own special circumstances, another researcher might very well have said that it was not TBI technology but immunology issues that were crucial and that she could not go on without dealing with them first.

The introduction of cobalt therapy was not enough, however, and the success that Thomas sought was still more than a decade away. By 1964, he had virtually given up on human studies and had turned, instead, to animal experiments. At a 1964 New York Academy of Science conference on TBI that primarily covered laboratory and clinical studies of radiation sickness, the noted radiation researcher Eugene Cronkite presented what was for all intents and purposes the consensus position of the attendees. Bone marrow transplants were, at best, a last resort for treating radiation injuries arising from total-body exposures. In his discussion, Cronkite referred to Thomas's work on treating leukemia with TBI as an important additional source of knowledge about treating radiation sickness. He argued that, at doses up to about four hundred rads, supportive therapy was highly effective in treating radiation sickness. Bone marrow transplant takes were possible, as Thomas had shown, but only at doses in excess of eight hundred rads, where the survival of patients or accident victims was, Cronkite argued, questionable and success was minimal or wholly absent. And, at intermediate doses, no therapy seemed to work.[37] As a means to treat radiation accidents, bone marrow transplantation was, Cronkite concluded, of little or no value. During the discussions that followed, Thomas could only dejectedly remark that his work supported Cronkite's conclusions: "We have attempted this procedure [TBI] and bone marrow transplantation in only two instances in the last year and a half." The first patient had died after only twelve days and the second after sixteen.[38] The treatment of both radiation sickness and leukemia with a bone marrow transplant had both reached an impasse.

The story of leukemia therapy took a better turn over the next seven years as research led to more powerful techniques for immunological (HLA) matching. In December 1971, Thomas set in motion a large-scale human study for treating leukemia with TBI, chemotherapy, and bone marrow transplants. He reported in 1977 that: "It is possible to achieve long-term remissions without maintenance therapy in some endstage patients with acute leukemia."[39] But the fatalities remained very high, owing, in part, to severe lung complications and the appearance of acute and chronic graft versus host disease (secondary syndrome). Thomas had modified the natural history of leukemia; the center of gravity of concerns had shifted from the recurrence of leukemia to the control and treatment of the substantial complications of the therapy. In spite of these difficulties, consensus on Thomas's technique was reached very quickly, without randomized trials, and has remained one of the mainstays of leukemia treatments, with various modifications, to this day. There is little question that his path to the 1977 paper was uncertain, messy, and fraught with contingent events (only some of which have been described here). It is also clear that the boundary separating Thomas and researchers studying radiation injury for military and industrial purposes was quite porous and that techniques and ideas crisscrossed between them. For example, the technique of bone marrow transplantation was first developed in the military laboratory, the discussion and classification of secondary syndrome in humans (graft versus host disease) had been identified and classified in studies on the radiation effects of nuclear war and industrial accidents, and an early platelet-producing unit for transplants had made its appearance in Cronkite's laboratory. At the same time, the efforts of Thomas and others in developing bone marrow transplants had been of great importance in trying to treat victims of radiation accidents in nuclear plants.

We have also seen that the close and enmeshed networks between research into leukemia and studies in radiation sickness were constantly changing. For example, Thomas initially argued that his program was as much about treating civilian populations following nuclear attack as it was about treating leukemia and other disseminated diseases. He moved readily between both areas, both drawing on and participating in conferences on military and industrial medicine as he grappled with a way to obtain consistent and extended relapse-free periods in his patients. Once he announced long-term survivors in 1977 and a consensus was reached in the medical community, the situation changed dramatically. TBI with bone marrow transplantation became a subdiscipline in its own right with its own special problems, methodologies, language, and acronyms. And, with these developments, TBI and bone marrow transplantation became divorced in later publications from Burchett's atomic plague and other cold war concerns. In a review article from the 1990s, to take one example, after recounting two examples of the injection of bone marrow prior to the Second World War, the authors go on to write: "The beginnings of modern BMT [bone marrow transplantation] may be traced to work showing that rodents can be protected against lethal hematopoetic injury by intravenous infusion of bone marrow. The subsequent identification of transplantation antigens . . . and the development of cryobiology . . . laid the groundwork for the difficult and time consuming clinical trials that brought allogeneic and autologous BMT to the present, albeit, imperfect state."[40] The one reference to 1950s research (the protection of rodents) scarcely provides a hint of the war-related studies that contributed to the present state of leukemia therapy.

Military Research with Cancer Patients

In the previous section, we saw how cancer therapy research with TBI and bone marrow transplant was fed by (and also contributed to) laboratory studies that were primarily concerned with diagnosing and treating radiation injury. In this section, I focus on research with cancer patients that was funded by the DOD to learn about the effects of radiation on military personnel. As we shall see, the ethical tensions between research and therapy that were evident in the DOD-funded studies were not unique to them but were shared by traditional studies such as cancer therapy clinical trials. Just as studies in cancer therapy and radiation injury traded in similar scientific ideas and practices, they also shared similar ethical practices.

In its study of human radiation experiments, ACHRE tied together the following elements to develop a history of human radiation studies for the military that used cancer patients. The committee argued that, during the war, concerns about the effects of radiation on workers at the Manhattan Project led to a series of TBI studies. The research was carried out at three academic medical centers, the Memorial Hospital in New York, the Chicago Tumor Institute, and the University of California Hospital. All the studies, except for a few of the cases in Chicago, were performed on sick patients and were meant to answer questions for the military about the effects of radiation on Manhattan Project workers. Approximately fifty individuals were entered into the studies, and the investigators claimed that the radiation treatments were a part of normal therapy, or they justified them with the argument that there was no other suitable treatment.[41] Saenger was, no doubt, a product of this tradition and was, likely, aware of the content of these wartime studies since they appeared in a volume on industrial medicine edited by Robert Stone with contributions from investigators at each of the three institutions that conducted wartime research on sick patients.[42]

Following the war, military interest in TBI resurfaced with the Nuclear Energy for the Propulsion of Aircraft (NEPA) project, where AEC and DOD planners sought to understand the acute effects of radiation on the crews of nuclear aircraft. The question of appropriate experimental subjects dogged early considerations by the DOD and the AEC about whether to fund the project. Initially, the head of NEPA's human experiments subcommittee, Robert Stone, argued that the studies should be performed on healthy volunteers, that patients were not a good substitute, and that the studies were possible on volunteers since the dose levels in the aircraft were not expected to exceed twenty five rads. A series of arguments ensued about the adverse publicity that might follow from using healthy subjects, about whether sick patients might be used instead, about whether the experiments should be classified, and so on. The planners and investigators also sought an ethical framework for the studies. There were, recall, no extant codes of practice at the time except for the AMA's three principles, which Stone used to support the research. The DOD, however, was still uncomfortable with the study and passed the final decision on to the AEC, which had, it argued, jurisdiction over civilian matters. By the time the program reached the AEC, it had expanded its purposes to also include research on the effects of radiation on military troops in a nuclear war. The AEC rejected the experiment on the grounds that there was sufficient information on radiation effects from the Japanese experience as well as the accidental exposures of a few civilians on the Manhattan Project.[43]

The NEPA project is interesting for a number of reasons. First, it clearly brings out the overlap between civilian and military research through the complex funding arrangements of the AEC. The AEC, remember, was a civilian agency that was created out of the Manhattan Project to maintain control of civilian and military use of atomic energy. It was later required to fund cancer research along with its many roles in controlling the distribution and use of atomic energy in medicine and industry. In the NEPA project, the AEC was involved in an ostensibly purely military program. During this same period, the AEC was also asked to fund what the M. D. Anderson Hospital and Tumor Institute believed would be the first high-energy cobalt-60 unit, which would presumably advance the treatment of all cancer with radiation. In these two projects, similar individuals were involved in funding decisions. In the case of the cobalt-60 unit, the AEC initially balked at supporting the project, and only later, under pressure from researchers at its own Oak Ridge Laboratory, did it contribute funding for the project. The cobalt-60 unit became an important resource for cancer studies at M. D. Anderson and led to attempts to improve most cancer therapy techniques, including large-field irradiation. The institution also sought and received military funding as part of its cancer therapy efforts, and, as we will see below, the cobalt-60 unit became a rationale for giving TBI therapy.[44] The mixing of military and cancer therapy funding and research is quite astonishing here. The AEC, an organization whose role was to oversee both military and civilian uses of nuclear energy, provided funds for a cobalt-60 unit for cancer therapy. M. D. Anderson used the acquisition of the machine as a means for eliciting DOD funds to carry out TBI experiments for the military. At the same time, the hospital also used the machine to develop TBI techniques for treating disseminated cancers that also had uses in its military program. It is no wonder, as we will see, that the research protocols for the military projects were so similar to cancer therapy protocols.

The second point is that healthy volunteers were still a potential source of research subjects at the time of the NEPA project. As we saw in chapter 2, during the early postwar period, experimentation with normal volunteer subjects was still being proposed. Sick patients, however, were viewed as participating in other types of experimentation; that is, they were more often than not identified with the small and measured modifications of therapy that presumably went on in a doctor's office. The AMA Code had been developed, like the Nuremberg Code, with healthy volunteers in mind. It was, therefore, apt for Stone to turn to it to support the NEPA studies. When the use of volunteers was rejected, he argued for cancer patients as a substitute. By the time of the later DOD studies at Cincinnati and elsewhere, volunteers were off the map, and no one any longer considered trying to enlist healthy subjects for radiation experiments.

Even though the AEC turned down the NEPA project, military planners sought other avenues since, by the beginning of the 1950s, there was increased concern about the possibility of atomic war with the Soviet Union. In 1950, immediately following the AEC decisions, the air force entered into a contract with M. D. Anderson Hospital to investigate the physical and psychomotor effects of TBI. Military research on radiation effects was also funded over the next decade at Memorial Hospital in New York, Baylor University in Texas, the Naval Hospital in Bethesda, Maryland, the AEC's hospital at Oak Ridge, Tennessee, and, finally, in 1960 at the University of Cincinnati. The Cincinnati studies extended until early 1972, when the president of the university ended its contract with the DOD, and, at that point, presumably all such military research had come to an end.[45]

All these projects were funded to investigate similar issues for the military: What would be the physical, psychological, and psychomotor effects of TBI on soldiers in a nuclear attack? Was it possible to find a biological marker in humans that would register the amount of radiation they had received? What type of therapy could be offered to soldiers who had received a sublethal dose of radiation? The last question, especially, was pursued vigorously by laboratory researchers on a number of fronts, as we saw in the last section. In these DOD investigations, the military was funding human rather than animal studies, and all the studies used sick patients, most with advanced cancers.

In its report, ACHRE linked these programs together since they shared some obvious characteristics. They all received military funding to answer military questions, and they all used sick patients. Moreover, they argued that the research was concerned, not with therapy for the patients, but with answering medical questions about military personnel under nuclear attack. According to ACHRE, the goal of answering military questions took precedence over the aim of treating the patients. I agree with that assessment in the Saenger case, but that conclusion is not at all unproblematic. Whether

Saenger favored research over clinical goals is one question. The fact that these different goals were in tension is quite another matter. However, an inference that should not be drawn is that tensions between research goals and therapeutic aims were peculiar to the DOD studies and that this in some way distinguished them from traditional cancer clinical trials. As ACHRE rightly pointed out, these conflicts are "still today open and vexing issues."[46] The DOD studies and traditional clinical trials shared too much for such distinctions to be made categorically.

To begin with, the DOD studies addressed medical problems about the diagnosis and treatment of radiation sickness that had applications to military personnel as well as civilian populations. The DOD-supported investigators were pursuing valid clinical questions, so it should not be surprising that the ideas and practices that they adopted were similar to, and, thus, not so readily distinguishable from, nonmilitary related research like cancer therapy.

In addition, the clinical ethics that informed the DOD studies shared much with clinical trials practices of the period. To take one example, the M. D. Anderson researchers designed their DOD project as a three-phase dose escalation trial. For the first and lowest dose regime (from fifteen to seventy-five rads), patients were entered into the study if they could still obtain "cure or definite palliation from established methods." As the dose was increased into the second regime (one hundred to two hundred rads), patients were treated if they could not expect significant benefit from procedures other than "systemic ones," while, in the third and highest dose regime (two hundred rads), "cure by conventional means was considered entirely hopeless."[47] One can find little substantive difference between the clinical ethos that informed this design and other non-DOD-supported clinical trials. Indeed, the researchers who designed the M. D. Anderson trial worked on a wide range of traditional cancer therapy trials and, thus, followed the standards of the time. One of the most notable design elements in this experiment was that, as the treatments became more toxic, patients had to be suffering from more advanced and intractable disease to be entered into the study, an ethical principle that was typical of its day and remains in effect to this day.

As I already mentioned, the argument that the military-funded studies should be clearly demarcated from traditional clinical studies because the military research aims were in conflict with the clinical goals is not at all straightforward. In fact, this argument consists of two separate but related parts: one is that there was a tension between research and therapy in DOD studies that set them off from traditional studies; and the other is that the

DOD studies used patients as proxies while traditional studies do not. The first argument, that DOD studies differed from traditional studies because of a tension between research and therapy, is difficult to support. Clinical trials throughout the postwar period were subject to criticisms about the conflict between research and therapeutic aims. Some medical researchers contested the ethical ground of randomized trials, claiming that the Hippocratic role of the physician was overridden by the demands of randomized trials.[48] The principle of equipoise, which provided physicians with an ethical justification to enter patients into randomized clinical trials, also came under specific attack.[49] The basis of the principle—that it was ethical to enroll patients when the treatment options were considered equally efficacious—meant equally efficacious for *cohorts* of patients who shared certain limited and well-defined disease criteria and other characteristics. But the ethical question that a physician confronted, the critics argued, was, not whether the arms of the study were in equipoise for a cohort of patients, but whether they were in equipoise for a particular patient with his or her own unique characteristics and problems. A physician, some critics contended, would almost always have made a judgment about which arm was best for a particular patient. Under such conditions, it would be unethical to enroll that patient in the randomized study.[50] If the physician did so—the argument would conclude—the research aims of the clinical trial took precedence over the interests of the patient. In this respect, the questions that arose for traditional clinical trials differed little from those that arose for the DOD-funded research in that there was an inevitable tension in the balance between research aims and clinical goals.

The second point, that the patients in the DOD studies were used as proxies while those in traditional studies were not, is also not as straightforward as it might seem. Of course, the patients in the DOD studies were proxies for soldiers, and the military goals clearly set these studies off from other clinical trials. But, as I discussed in the introduction, patients on clinical trials in a strict sense also always act as proxies for the cohort, although on traditional studies that may not always be so apparent to the patients or the investigators since the research aims are ostensibly about the patient's disease. In the case of military studies, the DOD investigators could not help but appreciate that they were using patients as proxies for the military and that there were conflicting goals between research and therapy. This should have given extra clarity to their clinical conduct, for example, that the therapy should in no way be compromised by the evident military goals.[51] The clear appearance of different goals gave them few grounds on which to later argue that they were unaware of the use of patients as proxies and the problems it could have for the patients' welfare. The problems for traditional cancer clinical trials were quite different, but more insidious, since the apparent equivalence between research and therapy goals might easily blind physicians into thinking that they were not using patients as proxies and that entry of the patient in a study was entirely for the patient's benefit. Indeed, the difficulty of realizing that patients in traditional clinical trials are proxies for the study cohort is nowhere more evident than when physicians apply the principle of equipoise for the cohort rather than for the individual patient. Although the DOD studies are clearly demarcated from traditional studies by the military component, in other ways they are not as distinct as we might at first believe. In both, the research goals and therapeutic aims are in tension, with the result that an individual patient's welfare could be compromised and, in both, the patients in a strict sense act as proxies for the study cohort.

These considerations should be kept in mind as we now turn to Saenger's TBI program. In one sense, this chapter provides us with a background to understand the importance of the paradigm problem of radiation injury and how it brought together a range of researchers (including Saenger) with interests in military and cancer medicine. Saenger's studies, as we shall see in detail in the coming chapters, could be located both in the cancer therapy and in the military research regions of the medical research environment. In another sense, this chapter and the preceding two provide us with a background in medical research and ethics that is grounded in the clinic and in the practices of the researchers of the first half of the postwar period. The case studies that I have followed—those of Bernard Fisher and Donnal Thomas—show a more complex picture of clinical trial practices than typical idealized accounts. In these studies, we have found that the directions of clinical trials are often modified, as exemplified by Thomas's change from military to leukemia therapy goals; that success is difficult to attain; and that social and political forces may often contribute, if not directly lead, to the pronouncement of the success of contested clinical trials, as in the case of Fisher and adjuvant chemotherapy for breast cancer. And we have also seen the highly contingent character, not only of medical research per se, but also of ethical regulation. The governance of clinical conduct did not so much evolve out of international ethical codes as it was the product of a political calculus of administrators at the NIH in the mid-1960s who understood that, if the medical community was to maintain control of its research destiny, it would need to govern investigators through an ethical review by their peers. And it was only after the mid-1960s that researchers like Bernard Fisher and Donnal Thomas had to follow NIH standards by introducing informed consent in spite of the fact that their clinical trials were highly toxic from their very beginnings. The vision presented in these first three chapters is one of a medical landscape filled with clinical trials with blurred boundaries between military and cancer research, where research is subject to changing goals as a consequence of contingent events and where ethical practice is but one of a host of issues that researchers need to negotiate in order to carry out their research programs. We should remember that it was in this landscape that Eugene Saenger was able to flourish.

Notes

1. One of the first techniques I learned as a medical physics trainee in radiation medicine in the early 1980s was how to treat painful widespread metastases with extended field radiation. The patients would be positioned on a mat on the floor of the treatment room, and partial-body (i.e., upper- or lower-body) radiation would be applied with a linear accelerator.
2. "Statement of Eugene L. Saenger, House Judiciary Committee, Subcommittee on Administrative Law and Government Relations," April 11, 1994, 4, DOD 042994-A;13/16.
3. Armitage et al., "Bone Marrow Transplantation," 295. The study in question is Thomas et al., "One Hundred Patients."
4. Dessauer, "Eine neue Anordnung."
5. Stone, *Industrial Medicine*, 4.
6. Oughterson and Warren, *Medical Effects*, 126.
7. Lindee, *Suffering Made Real*, 11.
8. I am using here a unit of dose that was more common during the period of Saenger's studies. At the time of the ABCC work, the distinction between exposure and dose was not so clear, and practitioners would often refer to them interchangeably with the designation r or R (for the Roentgen). In order to avoid unnecessary confusion, I use the rad as the unit of dose throughout. See, e.g., Johns and Cunningham, *Physics of Radiology*.
9. Warren, "Pathological Effects," 449, 450, 453.
10. Boyer, *By the Bomb's Early Light*, 331–53; and Winkler, *Life under a Cloud*, 84–108, 109–35.
11. Boyer, *Promises to Keep*, 30–31.
12. DOE, *Human Radiation Experiments*, 21, 27.
13. Cronkite and Breecher, "Whole Body Irradiation," 193.
14. Congdon, "Experimental Treatment," 746–51.
15. Cronkite and Breecher, "Whole Body Irradiation," 201–2.
16. Trentin, "Consequences," 212.
17. The use of *chimera* to describe a patient treated with a bone marrow transplant is troubling. The *Oxford English Dictionary* gives a number of definitions for *chimera*,

and most of them refer to its bizarre character. It is a "fabled fire-breathing monster" of Greek mythology with a "lion's head," a "goat's body," and a "serpent's tail" or a grotesque monster of incongruous composition and terrible character, "a horrible fear inspiring phantasm," an unreal creature of an "incongruous union of medley." Perhaps the medical use of the term was suggesting that some researchers felt a sense of having crossed the line in producing new creatures.

18. Trentin, "Consequences," 214.

19. Mathé, "Secondary Syndrome."

20. Barnes et al., "Murine Leukemia," 626.

21. Trentin, "Consequences," 214.

22. Thomas et al., "Intravenous Infusion," 491, 494.

23. Most of the studies from Hiroshima and Nagasaki had been kept confidential (see Day, "Medical Profession and Nuclear War"), and Ougherston and Warren's *Medical Effects* was published only in 1956.

24. Ewing, "Tissue Reactions," 113.

25. Sanderson, "Irradiation," 67. Sanderson was adapting a technique that was devised by Techendorff (see Techendorff, "Ueber Bestrahlung") and refined by Sluys (see Sluys, "Roentgenisation totale").

26. Medinger and Craver, "Total Body Irradition," 669.

27. One head and neck surgical technique of the 1950s characterized itself as a "commando operation" (*Lerner, Breast Cancer Wars*, 75). There are a number of critical studies of the increasing toxicity of modern therapies. The most famous, perhaps, is Ivan Illich's *Medical Nemesis*. See also the more measured, but still critical, Sharpe and Faden, *Medical Harm*.

28. Thomas et al., "Intravenous Infusion," 494.

29. Ibid., 492.

30. Ibid., 494.

31. Thomas et al., "Bone Marrow Transplantation."

32. Thomas et al., "One Hundred Patients," 512.

33. I am using rads here again, although the paper refers to "325r."

34. Thomas et al., "Irradiation of the Entire Body," 7, 12 (quote).

35. Ibid., 17, 16.

36. Ibid., 19.

37. Cronkite, "Human Radiation Injury," 345.

38. Thomas, "Observations," 394.

39. Thomas et al., "One Hundred Patients," 529.

40. Armitage et al., "Bone Marrow Transplantation," 295.

41. ACHRE, *Final Report*, 231–35.

42. Stone, *Industrial Medicine*. Reports on TBI used for the Plutonium Project were Craver, "Tolerance"; Low-Beer and Stone, "Hematological Studies"; and Nickson, "Blood Changes."

43. ACHRE, *Final Report*, 232–33. See also Whittemore, "Crystal Ball."

44. University of Texas, *First Twenty Years*, 212–23.

45. ACHRE, *Final Report*, chap. 8.

46. Ibid., 253.

47. Ibid., 236–37.

48. Hellman and Hellman, "Of Mice but Not Men," 1586.

49. For theoretical vs. clinical equipoise, see Freedman, "Equipoise," 143–45.

50. See the remarks in n. 29, chapter 1, above.

51. Some critics argue that human experimentation where research goals are unrelated to the disease of the patient is patently unethical. For a thoughtful presentation of this position, see Jonas, "Philosophical Reflections," 24–26.

PART II

A CASE OF CONTESTED KNOWLEDGE

4 Cancer Patients as Proxy Soldiers

Data derived from war and industrial casualties suffer because the exact amount of radiation received is unknown. In view of such ambiguities, observation of post-irradiation events in patients given total-body irradiation in known amounts seem pertinent. • William McFarland, "Hematological Events as Dosimeters in Human Total-Body Irradiation"

These studies are designed to obtain new information about the metabolic effects of total body and partial body irradiation so as to obtain a better understanding of these acute and subacute effects in human beings. This information is necessary to provide knowledge of combat effectiveness of troops and to develop additional methods of diagnosis, prognosis, prophylaxis and treatment of these injuries. • Eugene Saenger, *Progress Report*

....

In the late 1950s, Eugene Saenger launched a series of experiments, supported by the military, to study the effects of total-body irradiation (TBI). In his reports to the Department of Defense (DOD), he emphasized their value in providing knowledge about the combat effectiveness of troops on a nuclear battlefield. In other publications and venues, he also argued that his work provided important advances in the treatment of disseminated advanced cancer. In the next four chapters, I present a detailed history of Saenger's TBI experiments from their origin at the very end of the 1950s until their demise in 1972. The material in the earlier chapters offers a context in which to consider the ways in which Saenger's program shared in the practices of the period. The regimented yet contested character of clinical trials as well as the entrepreneurial methods of the investigators as they sought to enroll physicians, patients, and monetary and other resources are as evident in Saenger's program as they were in Bernard Fisher's chemotherapy studies. The relationships between cancer therapy and cold war military research to be found in Donnal Thomas's efforts were, if anything, even more apparent in the Cincinnati program. In a number of respects, the ethical practices of the Cincinnati physicians will also appear similar to those of their contemporaries, as will the influence on Saenger's research program of local peer review when it was introduced in Cincinnati following the surgeon general's notice of 1966 on the ethical review of research proposals.

But the local character of Saenger's program and the many contingencies that shaped it will also be apparent. For example, his military contacts during the Korean War had a profound effect on the initial direction of his TBI program. His collaborators, his medical peers, and the university administration also had various influences on the program, sustaining it, shifting its focus, and also contributing to its demise. In addition, Saenger's character, his brash and at times thoughtless behavior, had a telling effect, if nothing else, on the public scrutiny that was eventually brought to bear on his research program. In this chapter, I concentrate on that research program. By the mid-1960s, Saenger had initiated and greatly expanded his research efforts; in the late 1960s, he changed directions and added new initiatives. In the first two sections, I discuss in some detail how he was able to launch and build his TBI program using military support. His ability to take advantage of cold war issues and combine multiple goals clearly aligns him with many others in the research community. In the third section, I show how his strong clinical position within the medical center aided him in building a research enterprise. His power came in large part from his position as director of the Radioisotope Laboratory (RIL), which provided the space and rationale to maintain a number of interleaving programs as well as the patronage to enlist the many coworkers and other resources required.

In the last section, I discuss how Saenger used old and infirm patients as proxies for soldiers and why that was considered possible. I follow two particular examples: one using hematological measurements to indicate the radiation exposure of soldiers, the other using verbal samples as a measure of the cognitive response of commanders on the nuclear battlefield. I argue that an atomistic model of the body provided a rationale for Saenger to simultaneously treat patients and use them as proxies. On the one hand, their disease was the crucial reason for delivering and trying to elevate the dose of TBI. On the other hand, their sickness and debility were essentially bypassed when attempting to correlate the dose with the their metabolic and hematological measurements. In spite of Saenger's efforts, none of this research succeeded.

Establishment of a Research Program

Saenger's first research application—sent to the army surgeon general in September 1958—began innocuously enough with a brief discussion of what he termed *indications* in the literature of aminoaciduria following radiation. Saenger proposed to measure this effect, as well as mechanisms of immune response to TBI, in cancer patients. He suggested that, unlike the findings from nuclear reactor accidents, his would be the result of a controlled study that would include additional measurements like "amino acid excretion both before and after irradiation both to the whole body and parts of the body."[1] Saenger was suggesting that he would develop a human dosimeter by finding a marker like aminoaciduria that would register the amount of radiation an individual received.

The proposal received favorable responses from a number of military reviewers. One noted the study's "inestimable value in case of atomic disaster and nuclear warfare."[2] Another remarked on its great value to the military since, if Saenger succeeded, his approach "would lend itself easily to field testing as compared to some more exotic tests of bio-effects of radiation." That same reviewer also commented that Saenger's program would augment studies already in progress at Memorial Hospital in New York and at Baylor University, studies involving experimenters "working with humans."[3] Finally, the U.S. Army was pleased that the project had been crafted to fit in with its own immunological program, Saenger having proposed joint studies with investigators at Fort Knox.[4] Only one note of caution was raised, this by a reviewer who felt that the investigators would not succeed at correlating aminoaciduria with radiation dose. Yet he strongly recommended that the proposal should receive the highest priority since "there are so few radiologists who are willing to do total body radiation that those that are should be encouraged to do more." The investigators, he believed, would shift their focus once they realized that aminoaciduria would not provide a viable biological dosimeter.[5] By early 1960, the proposal was approved, and the DOD negotiated a contract with the University of Cincinnati School of Medicine.[6]

The proposal and its reception constitute a typical set piece illustrating how scientists adjust their arguments to the prevailing political forces and how they rely on a research ethos that is unarticulated but shared with their reviewers. This was clearly the case for Saenger's proposal, whose aims were strongly coded for its military readers. It is striking that Saenger made almost no claims about the military purposes, or, indeed, any purposes, of the study. The proposal was written as if he were engaging in a purely abstract problem—he would irradiate humans for no other reason than that there were "indications in the literature." Yet the reviewers understood his proposal and its value to the military, although each read his coded messages with a different emphasis. His work would be of value in the event of nuclear accidents, or for civil defense purposes, but mainly for the military's need for a simple method to measure radiation damage to soldiers on the battlefield during a nuclear attack. The use of cancer patients was also part of the shared code since Saenger made no attempt to justify under what conditions they might act as proxies for soldiers. He simply stated that he would irradiate cancer patients, as if their use were obvious. He assumed, rightly, that his reviewers understood that healthy volunteers were no longer politically viable, that patients were available to medical researchers as experimental material, and that they could readily be enlisted to serve the purposes of the study.

There was more to the coding of the proposal and its reception than simply a shared cultural understanding of medicine and military matters. Saenger had an inside track; he knew some of the players and what type of proposal they might fund. He would later remark: "I had been talking with a number of people in the Army, the Surgeon General's Office, and told them about my interest and so on, and they said they're interested in weapons effects."[7] His relationship with the surgeon general's office throws light on a surprising comment by one of the reviewers on the value of the proposal, namely, that "any correlation of tumor response to total dose of irradiation by means as proposed in this project would be of great value in the field of cancer."[8] Since there was no mention anywhere in the proposal of cancer research or therapy, much less of correlating tumor response with total-body doses, either Saenger had directly discussed his plans for cancer and military research with the reviewer, or they reached him through a mutual acquaintance. Whatever the path, it is clear that Saenger had a number of aims in mind, some of which he had discussed with individuals in the army. His goals included, not only military medicine, but also ideas about substituting TBI for chemotherapy in advanced cancer patients so that he could conduct cancer therapy and military research in one and the same patient.

Saenger's proposal also reflected his own research interests, which had evolved over the previous decade. Saenger credited his two years in the army during the Korean conflict as crucial to the development of his interests and abilities as a researcher in radiation science. He started his tour of duty at Sandia Laboratories, near Los Alamos, where, he recalled, "I got very good insight in bombology, because we were doing clinical radiology there." Sandia was involved in producing nuclear warheads, and there was much interest there in radiation effects, particularly from the fallout from the nuclear tests. Saenger took a number of courses at Sandia and developed an understanding of radiation physics. After six months, he was stationed at Brooke Army Hospital, Fort Sam Houston, as head of Nuclear Medicine. He became interested in the work of the surgical team in the burns unit, where isotope studies were needed. He became a consultant to the group and recalled that his work in the burns unit "taught him to think radiobiologically" and to appreciate clinical research. He had learned how to channel his growing interest in radiation effects toward research on humans since, as he later admitted, he "was never an animal person." He had also learned statistical methods so that when he "got into epidemiological studies . . . [it] all kind of fit together."[9]

When he returned in 1955 to the University of Cincinnati's RIL, Saenger was able to meld his interests in the long-term and acute effects of radiation and his background in radiation physics, human radiation biology, and statistical analysis. These interests were facets of a paradigm around which he could study any combination of the acute or chronic effects of radiation on either cancer patients or military or civilian populations. Even before submitting his proposal to the army surgeon general, Saenger had already been receiving support from the National Institutes of Health (NIH), to study the incidence of neoplasia in irradiated children, and from the Atomic Energy Commission (AEC), to prepare a handbook on the medical aspects of radiation accidents.[10] He was able to further extend his military and medical interests when he and a colleague in the Department of Pediatrics noted elevated levels of deoxyctidine in the blood of two patients treated for Hodgkin disease. According to Saenger: "We conceived in those patients . . . [that,] in place of giving chemicals to use whole body irradiation, we would look for . . . indicators of radiation injury."[11] He would substitute TBI for chemotherapy in patients with advanced cancers, and, at the same time, he would also look for a biological marker that indicated the amount of radiation received.

While Saenger was considering TBI as a means to do radiation studies, he was also aware that the radiotherapy program at the medical center would be able to carry out such investigations. The institution's first full-time radiotherapist had just arrived from Memorial Hospital in New York, an institution with a long history of TBI treatments. In addition, the hospital's first cobalt-60 radiation therapy unit had recently been installed and would provide an up-to-date machine for the project.[12] A number of disparate elements had come together at the end of the 1950s, and Saenger sought to synthesize the new staff, equipment, patients, and a credible research idea into a project that might be of interest to the U.S. military.

The timing was appropriate in another respect since the U.S. military's and Saenger's interests meshed by the end of the 1950s. During the 1950s, when Saenger was developing as a radiation researcher, the military was becoming increasingly concerned about nuclear warfare and the need for radiation studies. The Soviet Union had set off its first atomic bomb in 1949 and its first deliverable hydrogen bomb in 1955. Many in the medical profession responded to the growing nuclear fear of the time.[13] In an article in the *Annual Review of Medicine* reflecting the position of many of their peers, two prominent radiation researchers argued that their attitude on nuclear war had changed "from the defeatism of 1945–6 to one of constructive plans in preparation for atomic war if forced on the democratic world." Indeed, they implored the medical profession to "take an active role in the painful adjustment of society to the realities and potential acute and chronic hazards of nuclear explosions."[14]

One of the driving forces behind the military interest in radiation experiments was General James Cooney, who later (during the Korean conflict) recruited Saenger, bringing him to Sandia and Fort Sam Houston, and, presumably, taking him under his wing and greatly influencing him.[15] In 1952, when Cooney addressed a joint AEC/DOD panel on the medical aspects of atomic warfare, he declared that the "military has a definite problem . . . because an atomic bomb might be used as a tactical weapon." Under such conditions, a commander would be forced within a relatively short time to say to his medical officer: "I have 'X' thousands of men who have been subjected to various amounts of ionizing radiation from 25 to 150r or more. How many men can I take into battle? How many will be sick? When will they be sick? How many replacements shall I request and when shall I ask for them?"[16] Cooney's questions reveal the growing concerns at the time about the use of atomic weapons against American forces. In a 1953 report to Congress, the head of the AEC noted: "It is now possible to have a complete 'family' of atomic weapons, for use not only by strategic bombers, but also by ground support aircraft, armies and navies. The Department of Defense is very much aware of this change of concept, and atomic weapons are being incorporated into the operational plans of all three armed services." According to Secretary of State John Foster Dulles, these so-called tactical nuclear weapons "probably would be used if the United States became involved in any military action anywhere."[17] The notion that troops might very well get caught up in local tactical nuclear battles seemed quite possible.

The issues faced by various civilian and military groups were how to assess the amount of radiation received by individuals and how to triage and treat the victims—concerns that only heightened as the decade of the 1950s progressed. These issues were discussed at a 1957 meeting at the Pentagon among the heads of key government agencies.[18] All were to learn that there was "urgent necessity to obtain all possible biological and medical information" on the effects of nuclear weapons and that it was up to the medical profession to "fulfill adequately its mission and obligation to the military and the civilian population in the nuclear era."[19] Saenger submitted his proposal a year after this meeting, during which time he would learn about the heightened need for medical information on radiation effects through his contacts in the office of the army surgeon general. His proposal would also take on added importance since, a few months prior to submission, the AEC suffered a serious radiation accident at its Y-12 plant in Oak Ridge. The incident was widely known by radiation workers and was reported to the U.S. Congress. Saenger referred to the Y-12 accident in his proposal, suggesting that the data obtained from the victims (it seems that all five survived) were inadequate and that his study, which included pretreatment measurements, would provide superior data on radiation effects.[20] Saenger's interests and those of the military were on the same track by the late 1950s, and both sides sought out and embraced one another over the next decade.

Expanding the Research Program

Although Saenger's initial proposal appears to be limited in its goals, the intended scope of the program had appreciably widened by the end of the first year. By then, Saenger had treated seventeen patients (although he reported on only twelve), and he had already extended the search for a human dosimeter from measuring the level of amino acids in the urine to metabolic and hematological studies. He envisioned that the total-body doses might be increased to as much as six hundred rads, and he also planned to compare TBI with radiomimetic drugs. He also had the newly arrived cobalt-60 unit calibrated and a treatment technique devised for TBI.[21] A telling indicator of the extent of the program he had in mind was the number of investigators and the range of their responsibilities documented in the first progress report. There was Dr. Keriaekes in physics, who was responsible for dosimetry measurements, and Drs. Perry and Horowitz in radiation therapy, who treated the patients with TBI. In addition, Dr. Friedman carried out TBI laboratory studies and was responsible for patient selection (in collaboration with Perry and Horowitz). Alongside these major players, there were Drs. Berry and Guest in pediatrics carrying out metabolic studies, Dr. West in pediatrics and Dr. Luzzio at Fort Knox doing immunology, and Drs. Ross and Kaplan in psychiatry performing psychometric tests. Saenger had also obtained a bed in the psychometric ward until he could secure one in the soon-to-open metabolic unit.[22]

By the end of 1966, Saenger had treated forty patients.[23] By that point, the program had been substantially enlarged and the title of the project changed from the earlier "Metabolic Changes in Humans Following Total Body Radiation" to "Radiation Effects in Man: Manifestations and Therapeutic Efforts."[24] By 1969, the broad aim of the study had become "an investigation of critical organ systems like the bone-marrow and gastroenteric track."[25] This major shift to include therapeutic issues had occurred, in part, as a response to the clinical complications from higher dosages, complications that eventually led Saenger to investigate bone marrow transplantation. The appearance of toxicity in the patients at such an early stage had not been contemplated at the beginning of his studies, when he proposed developing a human dosimeter and measuring tolerance levels to TBI. With a bone marrow transplant, the patients would be changed, and these aims, as we shall see, would have to be adjusted. But Saenger could not continue his program without finding a way to sustain his patients, and, thus, he needed to adopt new strategies.

At the same time, the change of direction also reflected a new political environment at the University of Cincinnati College of Medicine. By the mid-1960s, Saenger's studies had come under the scrutiny of the Faculty Committee on Research (FCR), and he was forced to frame his research program to meet the expectations of his medical peers in addition to those of the military reviewers at the DOD. It is likely that the title of the 1966 proposal had been changed to include "Therapeutic Efforts" more to allay local concerns over the nature of the studies than to suggest changes in direction for the military. After the mid-1960s, Saenger explicitly began to include remarks in the proposals and progress reports on the value of his work for cancer therapy. In the 1969 proposal, for example, he noted that the Research Committee of the College of Medicine approved the study "in regard to its therapeutic value and informed consent." He also stated that the studies were performed on patients "as treatment for metastatic malignancy in place of chemotherapy" and that his data "suggest that [that treatment] may be equal or superior to current chemotherapy measures."[26] These statements were a far cry from his initial proposal, in which patients were barely mentioned; they were aimed at his clinical colleagues even though they were ostensibly written for military reviewers.

To expand his program to meet the demands of two such different professional groups—local physicians and military planners—Saenger had to walk a fine line. One strategy that he adopted was to frame his statements in such a way that they could be interpreted in alternative ways. For example, the title of his project, "Radiation Effects in Man: Manifestation and Therapeutic Efforts," could be interpreted by military reviewers to mean that this was a study of radiation effects in solders ("man") and how to treat their effects and by his medical colleagues to mean that this was a study of the value of TBI and bone marrow transplantation in treating cancer patients with disseminated disease. Saenger also coded his writing by using terms like *individuals*, *men*, or *humans* that could be read as meaning either "soldiers" or "patients," depending on the audience. He reserved expressions like *patients* for those times when he definitely wanted to situate the study as one in cancer therapy.

His rhetoric was not the only means by which Saenger tried to balance the demands of his military and medical peers. In the case of bone marrow transplantation, he attempted to develop techniques that would meet the specific needs of cancer therapy and military medicine at the same time. This strategy was a continuation of the earlier one where he appealed to both cancer and military communities by combining TBI for cancer therapy with the search for a human dosimeter. More specifically, Saenger proposed to investigate two techniques for bone marrow transplantation. In the autologous technique, the patient's own marrow would be extracted prior to TBI and then reinfused following the radiation treatment. In the homologous method, marrow from a healthy donor would be used instead. The first method had applications for the type of cancer patients he was treating, while the second was meant for battlefield applications. Saenger heralded the latter strategy as a "possible important future treatment method," one that would provide soldiers with temporary support while waiting "for the regeneration of the individual's own [blood forming] tissues."[27] But homologous transplants held greater dangers for the patients owing to possible immunological incompatibility between the donor and the host.[28] Saenger and his coworkers should have been aware of these problems since Mathé and others had published on the secondary syndrome well before Saenger submitted his initial proposal.[29]

From the mid-1960s, Saenger's program also began to actively incorporate studies of the cognitive and psychological effects of TBI on military personnel. These efforts were expanded in numerous ways throughout the latter half of 1960s, and they continued until the end of the program. Saenger even tried to align some of the cognitive investigations with the proposed transplant studies. He argued that he would look at the "decision making processes" (of soldiers on the battlefield), not only at three days, but also at six weeks postirradiation.[30] The latter date corresponded to the time when the bone marrow of a transplant recipient would have begun to recover.

Throughout the latter half of the program, Saenger continued to search for a human dosimeter, although the study of metabolic changes had given way to studies of radiation modifications in the DNA and RNA of blood elements. At the same time, he continued to reassure his military reviewers of the importance of his program. In the 1969 proposal, he declared that "there is no decrease . . . from the possibility of nuclear warfare," possibly to preempt any concerns that the disarmament talks at the time might diminish the "necessity to pursue . . . investigations of acute radiation effects and the attendant treatment possibilities in the human being."[31]

This brief review of the trajectory of Saenger's program highlights his ability to adjust the program's aims in response to contingencies and his expansion of the military goals and incorporation of a new therapeutic profile with bone marrow transplants. The latter was a response to clinical problems as well as the changing ethical environment. As a consequence, Saenger was faced with difficult challenges as he sought to satisfy his medical colleagues on the local peer review committee. These challenges will be examined in some detail in chapter 6, but first I situate Saenger's research program more firmly in the landscape of the College of Medicine.

Situating the Research Program: The RIL

The TBI research effort thrived, in part, because it was under the umbrella of the RIL. Saenger had set up the RIL in 1949 as an assistant professor in the College of Medicine, but he did not begin to develop it until he had returned from military service in 1955.[32] The laboratory grew rapidly from one primarily providing diagnostic services with radioisotopes to one encompassing a range of clinical, research, training, and radiation health-related services. Through an NIH grant, Saenger was able to create a training program that included not only courses in nuclear medicine (isotope studies)—the nominal purpose of the laboratory—but also those in radiobiology, radiotherapy, and radiological physics. The grant also provided the funds to hire a chief of radiation physics, which had important implications for the authority of the laboratory.[33] In addition, since Saenger also maintained therapeutic isotope procedures within the RIL, his program extended into radiotherapy practices.

The scope of Saenger's influence through the RIL spread even further than these clinical efforts suggest since there were also a number of research programs under the RIL's purview. A 1965 table of organization of the RIL research activities contains a total of nineteen projects in the areas of nuclear medicine, radiobiology, radiotherapy/cancer chemotherapy, computer applications, epidemiology, and radiological physics.[34] Significantly, the TBI research was placed under radiobiology rather than radiation therapy. Perhaps Saenger considered the TBI effort as predominantly biological research supported by "conventional" TBI therapy (as he later claimed), or perhaps he considered the research with bone marrow transplantation too uncertain to list the project under therapy research. The TBI program itself was divided into three subprojects: total-body treatments under the direction of Saenger, bone marrow transplantation under Friedman, and metabolic studies under Berry.[35]

Not unlike other hospital facilities that combine clinical and research duties, the RIL required various personnel to operate, measure, and interpret the output of various therapeutic and diagnostic devices located both within the RIL (e.g., isotope studies of bone metastases) and at other sites (e.g., TBI in radiation therapy). Consequently, the RIL contained under its umbrella physicians, nurses, technicians, physicists, computer programmers, statisticians, and other staff who were recruited for the different projects from various departments and divisions within and beyond the hospital. The laboratory's efforts extended throughout the hospital, from radiotherapy to X-ray diagnosis, pediatrics, hematology, the psychometric and metabolic wards, the computer facility, and so on.

Each of the sites in this network fed the RIL clinical reports, radiotherapy records with patient dosages, hematological readings on specially designed forms, X-ray films and diagnostic reports, photomicrographs of chromosome breakage, and so on. The data were decoded and transformed into a unified language, entered into computers, and further reduced and analyzed. For example, Saenger's efforts to correlate radiation dosages with hematological profiles were typical of such practices. The RIL was, according to a now familiar paradigm, a center of calculation for this large and complex network[36] where Saenger tried to construct scientific knowledge using data sets transported from diverse locations. These far-flung sources of data were held together by a system of metrology for establishing and maintaining standards, operated by the RIL. This metrology included the calibration of devices and operational rules of practice that ensured that measurements made in different locations were commensurable. For example, calibrating the cobalt-60 machine and enforcing clinical protocols for treating the patients maintained the TBI dosages.

Such centers of calculation and metrology were discussed in chapter 1, where it was noted how the paradigm fit the multicenter clinical trial, under which protocols emanating from a coordinating center (statistical unit) were maintained at the participating hospitals through continuing oversight (quality control). This produced (in principle) a static but flexible metrology for the limited period during which the trial operated. The RIL also operated as a center of calculation; its maintenance of standards had to be dynamic since the programs were exploratory and the measurements constantly modified in response to new findings.

Saenger's far-flung research enterprises were possible in great part because of the strength of the RIL. The clinical responsibilities of the laboratory meant that Saenger was able to amass considerable power within the institution. He was able to build up an infrastructure that provided equipment, staff, and expertise that could be drawn on to help develop and carry out the numerous research projects, which, in turn, provided additional staff, equipment, and funds. The combination of clinical and research programs provided Saenger with patronage considerable enough to engage other departments to participate in his programs.

At the heart of this diverse enterprise were the patients whose need for clinical service was the basis for the RIL. When the dean of the College of Medicine attempted to move some of the clinical isotope studies out of his laboratory, Saenger responded immediately and presciently. One of the purposes of putting isotope studies in the lab, he argued, was to provide "clinical material" for the residency training program. But, most important, the transfer of isotope studies would change the lab so profoundly that "many of these programs would not be either medically or scientifically acceptable."[37] Saenger was arguing that the clinical and research programs were so intrinsically tied together that one could not exist without the other. If he lost the patients, his revenues would be lost, equipment would go elsewhere, the training program would become a shambles, research support would dry up, and his entire operation would collapse like a house of cards. In this sense, it was misleading when, after later public disclosures, Saenger and the university argued that the money from the DOD was used only to support the staff and diagnostic tests specifically required for the TBI study. The hospital, the College of Medicine, and the DOD were involved in a much more symbiotic relationship. Military funds were used to develop TBI techniques that were used by the medical center, while clinical services, supported by public and private funds, were exploited for the DOD research projects.

Cancer Patients as Proxy Soldiers

In his studies, Saenger used sick and infirm patients as proxies for soldiers. Why were such patients even considered as possible surrogates for healthy soldiers? On the face of it, the prospects for success should have appeared hopeless. The first group of patients treated, for example, averaged almost sixty years of age; all were very ill and appear from the clinical reports to be not at all like the patients in the "good nutritional state" that Saenger had described in his proposal.[38] Yet the use of such patients as proxies did not raise questions from his contemporary reviewers. No one questioned the issue of proxy per se, in part, I believe, because of a shared conception of what it meant to conduct cancer research. To begin with, Saenger's peers took for granted that patients with the most advanced cancers were (and still are) the most appropriate subjects for the most exploratory trials testing cancer treatments. Moreover, a medical model of the atomized body provided a rationale for using the responses of the patients' organs and tissues as surrogate measures for those of soldiers' organs and tissues.[39] The patients were merely instruments or vessels who mediated between the dosage levels of TBI that they received and the responses of their organs. Without such a tacit understanding of clinical research, it would have appeared utterly ludicrous to use old and ill patients as stand-ins for young and vigorous men at war.

Experimenting on such patients was something like tuning a radio. Andrew Pickering argues that tuning is a goal-oriented practice in which scientists, who are "human agents in a field of material agency," construct experimental devices to manipulate those material agencies and monitor their response. If the investigators are not satisfied, they adjust the experiment, monitor it again, and so on—in a sequential "dialectic of resistance and accommodation"—until the process locks in and a stable and repeatable condition is achieved.[40] Pickering's examples are taken from physics and technology, although he notes that the tuning metaphor was used earlier in a biological setting by Ludwik Fleck to describe the development of the Wasserman reaction.[41] The tuning metaphor can also be applied to human experiments since interventions are made, responses are monitored, and a stable correlation is sought between the interventions and the responses.

In this sense, tuning would, in Saenger's experiments, mean adjusting the radiation doses and trying to correlate them with the various metabolic and hematological parameters measured. Here, the patients as individuals are abstracted away, becoming little more than conduits to their organs, tissues, cells, and molecules, the focus of Saenger's experimental manipulations. The patients were envisioned as doing no more than mediating between the cobalt-60 machine and the various monitoring devices that provided data for analysis.[42] In this view of medical research, it appears to have made little difference whether the patients were sick or healthy, young or old. To be sure, their age and disease status might influence the response of their organs. But, in such cases, the effects would be considered secondary, and, if necessary, the patient-specific effects could be determined and subtracted away, leaving a universal measure of radiation response.

The tuning analogy applies in another sense to Saenger's experiments. Saenger was constantly adjusting his patient population so that he could deliver the doses he required. He was, in effect, modifying his experimental apparatus (consisting of patients, cobalt unit, and protocols), and he needed the appropriate patients to make the system work, that is, to deliver the TBI doses he required. It would not be far-fetched, then, to suggest that Saenger's use of bone marrow transplants was an attempt to repair his experimental apparatus by making his patients more robust to the TBI they received. The mechanistic language may seem a bit heavy-handed; I could have used the expression *experimental system* instead of *experimental apparatus*. But the mechanistic language provides further insight into the fact that Saenger could take contradictory positions on the state of the health and disease of the patients. On the one hand, the status of the patients had little relevance for correlating dose with hematological and metabolic measurements. On the other hand, it was crucial to gauging increases in the dose of TBI. Only a mechanized model could support both points of view. To appreciate these studies, I follow Saenger's attempts at finding a population of patients to which he could deliver the radiation doses he required to construct a human dosimeter. I also follow two of his efforts at correlating doses: one with hematological profiles, the other with verbal measures of cognitive ability.

In his initial proposal, Saenger argued that he could correlate aminoaciduria in patients with increasing doses of TBI. Since the patients were meant to represent soldiers on a nuclear battlefield, he proposed to tune the experiment by using only adult males with proven metastatic cancer and in "good nutritional state." He tuned out women since, he claimed, aminoaciduria was sensitive to their menstrual cycle. He also eliminated patients with lymphomas since the rapid response of their tumors to radiation would likely confound his measurements.[43] By the end of the first year, Saenger had further adjusted the patient population. His first year's report indicated that five of the first twelve patients he treated were female. Although he did not discuss why he was including women in the experiments, we can assume that he was finding it difficult to recruit patients. Besides, he had turned to other measures of radiation response for which the gender of the patients would have less importance.

Saenger was also developing methods by which he could account for the disease status of his patients. Among the new parameters of response was the so-called hematological profile or score, which, he argued, would provide "a way of assigning rank score for deviations from normality." These profiles had been developed for monitoring the responses of nuclear power plant personnel to nuclear accidents and were combinations of approximately 15 different blood counts ranging from red cells to white cells to platelets and neutrophils. Saenger found that "most of the patients had positive scores prior to radiation" which meant that the profiles as they stood were not proxies for a normal population of soldiers whose scores should be at or near zero. He interpreted this pretreatment profile as a "manifestation of neoplastic disease," that is, as a measure of the effect of cancer on the measured hematological profile. He also assumed that the pretreatment profile did not change over the period that he treated and monitored the patients.[44] The assumption was not unreasonable since he was treating patients with slowly responding tumors. If this were the case, then the pretreatment profile could be subtracted from a post-therapy profile to produce one that indicated the effect of radiation alone. The net score, then, with the pretreatment neoplastic component subtracted, could be translated, Saenger believed, onto a universal scale that he would apply to soldiers. Nevertheless, he was never able to obtain a stable correlation between radiation dose and a number of alternative hematological profiles.

Although Saenger felt that he could subtract the effects of disease, infirmity, and age using the pretreatment profile, he was, nonetheless, having substantial problems delivering higher doses. One patient with carcinoma of the left lung who had received 150 rads of TBI had died on the thirty-fourth day after suffering from anorexia, malaise, weight loss, leucopoenia, thrombocytopenia, and anemia. A second patient with renal disease died within thirty-seven days and a third after ten months. Since many of them had also received prior radiation or chemotherapy, this made them, Saenger believed, more sensitive to TBI. He changed the entrance requirements by eliminating patients with such characteristics in the hope that the remaining patients would be more robust and that he could then go on to higher doses. In the meantime, he reported that his data showed that the combat effectiveness of soldiers would begin to deteriorate rapidly when they were exposed to doses rising above 150 rads. He also claimed that soldiers would have more pronounced responses to nuclear attack if they were previously exposed (a position that he would later reverse).[45]

In spite of his attempts to exclude previously irradiated patients, Saenger had to finally concede that the disease status and overall health of his patients did matter. He realized that, "to proceed with higher doses, we feel the need to protect our patients even if we might sacrifice their value for hematological evaluation after 2–3 weeks."[46] He began to use bone marrow transplantation to counter the complications arising from the higher doses. In this process, he manipulated the patient's bone marrow, which, within weeks, might produce new blood elements and, most important, more stable patients. But he recognized that, because he manipulated the patients, he could no longer search for tolerance levels and that it would also not be possible to find a meaningful correlation between a hematological profile and dose. Nevertheless, he could look for other candidates for surrogate dosimeters that did not use blood products and might not be affected by the transplant. In addition, bone marrow transplantation was a therapeutic tool that could be used to treat soldiers in the field or advanced cancer in the clinic.

Although Saenger had difficulty increasing the dose and in finding a human dosimeter, he still sought to correlate other responses to TBI, especially in a series of cognitive and psychological studies. In these cases, as elsewhere, the belief that the patient's status could be ignored or corrected for was powerfully embedded in his thinking. Nowhere else was the atomized body more chillingly exploited than in the cognitive and psychological studies performed throughout the project. By 1966, Saenger's team began to measure what were termed *performance decrements*, which, he claimed, would "provide information in another important parameter of combat effectiveness of troops."[47] He ruled out treadmill studies and other measures of physical work as untenable since the "patients are old and ill and may have unknown metastases." But he could, he claimed, investigate the effects of TBI on patients using measures of repetitive work to provide a proxy for a soldier driving a jeep or using cognitive measures that would act as a surrogate for commanders making decisions on the field of battle.[48]

Saenger's coworkers developed a range of measures of performance decrements and of the psychological effects of radiation on the patients. One of the psychological measures, the so-called Halstead Impairment Index, was believed to be very sensitive to "organic states of the brain." Saenger reported: "The results thus far suggest that all patients in their baseline measures demonstrated cerebral organ deficit."[49] To put it crudely, he was claiming that all the patients were retarded! In spite of this claim, he suggested that he still might develop a workable measure. He had in mind something along the lines of separating out the "organ deficit," that is, subtracting the supposed retardation of his patients, to obtain a surrogate measure of a "mentally normal" individual's response to radiation.[50]

The investigators had great difficulty with the debilitated state of the patients. Yet they continued to develop new methods and turned increasingly to measuring cognitive impairment on the basis of various types of "content analysis of verbal behavior."[51] The situation was summed up in the 1968–69 progress report:

> The physical condition of an overwhelming proportion of the patients seen for psychological evaluation over the course of the last five years has been such that they had been unable to undertake even the most simple performance tests with any consistency. At times, this has been due to difficulties with vision, use of hands or total physical disability which precluded sitting up in bed. In other instances the low level of basic intellectual functioning of the patient has precluded adequate task performance. We have therefore been forced to rely in large measure on the effect of radiation treatment on cognitive functioning via the content analysis of verbal behavior.[52]

The investigators were forced to use verbal samples since those samples were the only data that they could possibly extract from the patients. Yet they persisted in believing that verbal scores could be reliably captured and that they could predict the cognitive effects of radiation on soldiers and commanders in the field.

Saenger believed that he could map the clinic onto the nuclear battlefield. His patients would act as proxies for soldiers, while he would merely gather data, without intervening. Yet he actively recruited only certain types of patients, delivered radiation according to protocols that were (at least in part) influenced by military questions, and modified his patients with bone marrow transplants. He took the prevailing experimental ethos to support a view according to which atomized patients passively yielded data for analysis while the research had no influence on the quality of the therapy. During an unguarded moment, Saenger may have acknowledged that this was not the case, and that he was manipulating the patients, when he remarked: "Somehow, we could never *get* the expression [of a dosimeter] in patients whom we treated" (emphasis added).[53] He was not so much trying to *find* a biological marker to register radiation as he was attempting to *construct* a human dosimeter.

Notes

....

1. Eugene L. Saenger, "Metabolic Changes in Humans," *Contract Proposal*, September 25, 1958, 3, DOD 042994-A;1/16.

2. Isherwood to Ass[istan]t Ch[ief], Biophysics and Astronautics, Memo, October 22, 1958, DOD 042994-A;15/16.

3. Sullivan to Hullinghorst, Memo, November 13, 1958, DOD 042994-A;15/16.

4. Cox to Sullivan, Memo, November 3, 1958, DOD 042994-A;15/16.

5. Hartering to Sullivan, Memo, November 7, 1958, DOD 042994-A;15/16.

6. Eugene L. Saenger, "Metabolic Changes in Humans Following Total Body Radiation," *Negotiated Contract Number DA-49-146-XZ-029*, January 1, 1960, DOD 042994-A; 15/16.

7. Ron Neumann, Gary Stern, and Gil Wittemore, Interview with Dr. Eugene Saenger, Cincinnati, September 15, 1994, 43, ACHRE Interview Project (hereafter cited as Saenger Interview 1).

8. Isherwood to Ass[istan]t Ch[ief], Biophysics and Astronautics, Memo, October 22, 1958, DOD 042994-A;15/16.

9. Saenger Interview 1, 4, 14.

10. This handbook was Saenger's 1963 edited volume in which the wartime studies of radiation effects on patients had been documented (see Saenger, *Medical Aspects*).

11. Saenger Interview 1, 43.

12. "Statement of Representative John Bryant, House Judiciary Committee, Subcommittee on Administrative Law and Government Relations," April 11, 1994, 5, DOD 042994-A;13/16.

13. Typical of such considerations is the following statement by Jacobson et al.: "The experiences gained at Hiroshima and Nagasaki have established certain facts which permit the formulation of certain broad principles to be followed if large cities are bombed" ("Physicians and Atomic War," 138). See also Boyer, "Physicians Confront the Apocalypse"; and Day, "Medical Profession and Nuclear War."

14. Cronkite and Breecher, "Whole Body Irradiation," 207, 193.

15. Saenger Interview 1, 5. As early as 1949, Cooney had written on the psychological effects of atomic warfare (see Cooney, "Psychological Factors"), and, in 1951, he had addressed the Radiological Society of North America on issues related to the effects of atomic warfare on civilian populations (see Cooney, "Physician's Problem").

16. Joint Panel on Medical Aspects of Atomic Warfare, Minutes of Seventh Meeting, January 25–26, 1952, 3, http://search.dis.anl.gov/ (accessed July 2001; hard copy in author's files).

17. Both AEC head and Dulles quoted in Kornfeld, "Nuclear Weapons," 209.

18. These included the directors of the AEC's Division of Biology and Medicine, Civil Defense, and Armed Forces Special Weapons, the secretary of the Department of Health, Education and Welfare, and the surgeons general of the three armed forces. Armed Forces Special Weapons was soon to become the Defense Atomic Support Agency, which funded Saenger's first proposal. It later became the Defense Nuclear Agency. In the text, I do not distinguish between them and use the more general term *Department of Defense*.

19. Department of Defense, Intra-Agency Conference on Biomedical Effects of Nuclear Weapons, Minutes Pentagon, Washington, DC, October 9–10, 1957, 3, 2, DOD 112194-A;3.

20. Eugene L. Saenger, "Metabolic Changes in Humans," *Contract Proposal*, September 25, 1958, 3, DOD 042994-A;1/16.

21. Eugene L. Saenger, "Metabolic Changes in Humans Following Total Body Radiation," *Progress Report*, November 1961–April 1963, 31–46, 1, 13–19, DOD 042994-A;5/16. Saenger had treated two patients on an orthovoltage X-ray machine.

22. Ibid., 2.

23. Eugene L. Saenger, "Metabolic Changes in Humans Following Total Body Radiation," *Progress Report*, February 1960–April 1966, 1, DOD 042994-A;7/16.

24. Eugene L. Saenger, "Radiation Effects in Man: Manifestations and Therapeutic Efforts," *Progress Report*, May 1967–April 1968, 1, DOD 042994-A;5/16.

25. Eugene L. Saenger, "Radiation Effects in Man: Manifestations and Therapeutic Efforts," *Contract Proposal*, February 19, 1969, 1, DOD 050294-A;1/6.

26. Ibid., 5.

27. Ibid., 4.

28. I could find no indication that Saenger and his team ever carried out homologous transplants.

29. Mathé, "Secondary Syndrome."

30. Eugene L. Saenger, "Metabolic Changes in Humans," *Contract Proposal*, September 25, 1958, 4, DOD 042994-A;1/16.

31. Ibid.

32. Eugene L. Saenger, Curriculum Vitae, DOD 042994-A;13/16.

33. Radiological physics provided services to the three divisions within the Department of Radiology—Nuclear Medicine (diagnostic isotope studies), Radiology (diagnostic X-ray studies), and Radiotherapy—and having the director of physics under the RIL gave Saenger influence over operations within other divisions.

34. Radioisotope Laboratory University of Cincinnati College of Medicine, Table of Activities, 1965, DOD 042994-A;1/16.

35. Some of the names in the table were placed in parentheses, which may have indicated that the individuals in question were not directly within the reporting structure of the RIL. If that was the case, then Saenger had all physics personnel under his purview, which would have given him substantial power in radiotherapy operations.

36. Latour, *Science in Action*, chap. 6.

37. Saenger to Grulee, Memo, April 23, 1965, DOD 042994-A;1/16.

38. Eugene L. Saenger, "Metabolic Changes in Humans Following Total Body Radiation," *Progress Report*, February 1960–October 1966, Clinical Profiles, DOD 042294-A;7/16.

39. Löwy, "Experimental Body," 440–43.

40. Pickering, *Mangle of Practice*, 21–22.

41. Fleck, *Genesis and Development*, 97.

42. The same idea of tuning can also be applied to grouped cohorts in clinical trials where the end points are the survival of the patients and not the responses of organs and tissues. In that case, stratified samples are created (subgroups of patients with

such different characteristics as tumor stage, nodal status, gender, age, etc.). Each subgroup represents one tuning of the experiment, and a stable correlation for each subgroup is sought between the different treatment arms and survival.

43. ACHRE's most damaging claim against Saenger is that he chose only patients with slowly growing (radioresistant) tumors since the slow response of their tumors to radiation would not interfere with the metabolic measurements he needed for the military aims of his research. ACHRE further argued that the doses of TBI that Saenger used were too low to affect such tumors; consequently, without bone marrow transplants to permit him to deliver higher doses, his treatment strategy was unethical. By the mid-1960s, however, Saenger had begun to develop bone marrow transplants to be used with TBI. ACHRE's ethical argument then turns, to some extent, on a technical point, namely, when consensus was reached in the community that TBI without a bone marrow transplant was no longer a viable therapeutic option (ACHRE, *Final Report*, 251). See also chapter 8.

44. Eugene L. Saenger, "Metabolic Changes in Humans Following Total Body Radiation," *Progress Report*, February 1960–April 1961, 20, 21, 61, DOD 0050294-A;2/6.

45. Eugene L. Saenger, "Metabolic Changes in Humans Following Total Body Radiation," *Progress Report*, November 1961–April 1963, 7, 18, DOD 042994-A;5/16.

46. Ibid., 17.

47. Eugene L. Saenger, "Metabolic Changes in Humans Following Total Body Radiation," *Progress Report*, February 1960–April 1966, 2, DOD 042994-A;7/16.

48. Eugene L. Saenger, "Metabolic Changes in Humans Following Total Body Radiation," *Progress Report*, February 1960–October 1966, 9, DOD 042294-A;7/16.

49. Ibid., 11, 15 (quotes).

50. Eugene L. Saenger, "Radiation Effects in Man: Manifestations and Therapeutic Efforts," *Progress Report*, May 1968–April 1969, 32, DOD 042994-A;2/16.

51. Eugene L, Saenger, "Metabolic Changes in Humans Following Total Body Irradiation," *Progress Report*, May 1966–April 1967, 28, DOD 042994-A;2/16.

52. Eugene L. Saenger, "Radiation Effects in Man: Manifestations and Therapeutic Efforts," *Progress Report*, May 1968–April 1969, 28, DOD 042994-A;2/16.

53. Saenger Interview 1, 37.

5 A Cancer Patient's Story

Come, Jesu, come, my flesh is weary,
my strength is fading more and more,
I long for thy peace;
the bitter path grows too hard for me!
J. S. Bach, Motet, BWV 229

The analysis of our 88 treated patients shows that 44 per cent experienced no symptoms at all, and that 27 per cent had transient nausea and vomiting within 3 hours, and 14 percent within 6 hours and 3 percent within 13 hours. In only 4 patients (4 per cent) were the nausea and vomiting of a severe nature. These symptoms are no greater than found after surgery or after treatment with cancer chemotherapy drugs. · Saenger, "Whole Body and Partial Body Radiotherapy of Advanced Cancer"

Saenger and his coworkers viewed patients with advanced cancers in various ways: as patients who needed treatment, as surrogates for soldiers, as part of the experimental apparatus, and as patients "beyond cure" and, thus, appropriate for exploratory medical studies. In these and later depictions, the patients' stories are narrated by the researchers, the University of Cincinnati administrators, and various critics. The patients themselves are abstracted and appear to be little more than objects caught up in a medical research system. There is some truth to this depiction. At the same time, we lose sight of the fact that they were also individuals who acted on the world, who tried to play out, as best they could, the bad hand that life had dealt them.

The sketch given of Saenger's program in the previous chapter is sufficient to allow me to tell his patients' stories from their perspective in this chapter, to try to give some idea of what it may have been like for them to confront their diseases and participate in the total-body irradiation (TBI) experiments. I do so because such stories can, when kept in mind, serve as a corrective to the representation of the instrumental role that Saenger's patients will continue to play in later chapters. Their voices will remain audible and not be drowned out by the din of medical and social confrontations.

In the chapter, I tell just one story, that of Maude Jacobs, patient 45, who received TBI in November 1964 and died later that year. If I were trying to write a social history, it might be important to tell a number of stories and to know the distributions of the patients according to class, race, age, survival, complications, and other factors. In such a compilation, the patients' voices would, however, once again be lost through their presentation as members of various categories, this time social historical ones. Rather, I want to bring out one patient's experience as an individual story, something that can be set against the narratives of the researchers, the hospital administration, and various critics that too often obliterate the patients' voices.[1] To do so, I present in detail the difficulties and the contingencies of one patient's encounter with disease, the physicians, the hospital, the therapy, and family and friends. It does not matter very much for this story whether Maude Jacobs was like or unlike her compatriots: indeed, in many ways she was anything but typical. She was younger than many of the others, she had one of the shortest survival periods following TBI, she did not go through a bone marrow transplant, nor was she given any of the psychological tests. Yet her overall experiences exemplify many of the situations the patients faced. She was terrified by the prospect of cancer, there was every reason for denial, and she agreed to TBI because of the high repute in which she held her physicians. And, most important, because the procedure had been standardized—all the patients were led into the same room, and all of them underwent a similar protocol—Maude's experience of TBI must have been similar to the others'.

There are also some practical problems that would make it difficult to tell many of the individual stories. The materials needed to reconstruct the case histories are very limited. There are, of course, surviving hospital records, but they have huge gaps in them, sometimes at the most critical junctures. For instance, there is nothing available for Maude's crucial hospitalization following TBI. In addition, the records are in terrible condition since they were stored on microfiche and later (in the 1990s) photocopied. Even when they have been reasonably well preserved, many of them are written in longhand. The handwriting is usually so poor that it is often difficult, if not impossible, to reconstruct what an administrator, a physician, or one of the other hospital workers meant to convey. Dates on notes are rare, and initials (there are almost never signatures) are almost always indecipherable, if they appear at all. (During the Advisory Committee on

Human Radiation Experiments [ACHRE] hearings in 1994, as we will see, the families of the patients complained bitterly about the state of the hospital records.) And, of course, the hospital records were written, not by the patients, but by hospital workers who presented the patients' cases in highly technical language, representing them more as medical objects than as human subjects. In spite of these difficulties, enough can be extracted from the charts, Saenger's reports and publications, and what we know about radiotherapy and hospital practices of the day to imaginatively reconstruct bits and pieces of Maude Jacobs's treatment experience. Also extant are some contemporary letters from Maude to her aunt during a crucial phase of her disease and some later testimony by her children. Although the documentation is still spotty, a plausible portrayal of some events in Maude's final months can be attempted.[2]

First Encounter with Cancer

Maude Jacobs relates a curious fact to her interviewer at Cincinnati General Hospital (CGH) in July 1964. She had fallen some eighteen months previously and injured her right breast, arm, and back. She felt a swollen area on the upper portion of her breast that did not go away and, in fact, began to grow. Maude then recalls another defining moment, about one year after her fall, when her lower back and her right leg began to bother her. These two events are recorded on her admission note for CGH: the first denotes the onset of a problem—a moment that Maude identifies as the cause of her difficulties—and the second is linked to the first, possibly by Maude, and certainly by the interviewer who considered the possibility of metastatic disease. The second event is the more crucial for Maude since it is what has brought her to CGH. It marks the moment when pain was no longer a stranger, coming to visit every so often, and then disappearing from consciousness, but a constant inhabitant of her body. Maude complains that the pain in her back is aggravated when she moves around or does any work at home. Perhaps it has already changed her daily routine, causing her to rest now and then or to not take on certain tasks that she unconsciously realizes may cause further discomfort. It is not just her back that bothers her, she tells the interviewer; the pain extends along her right side from her thigh to her knee and all the way down to her ankle.[3]

On the following day, Maude went to the X-Ray Department to have "pictures" taken. It is possible that all she remembered of the examinations was the likely repeated admonitions by the technicians to "hold still" before each exposure. We can be certain that she did not see the X-ray report, which began cautiously, suggesting that she did not have metastatic disease in her lungs since "the lung fields appear to be free of infiltrate." The rest of the report was anything but positive, and the bone survey was clearly foreboding: "There is increased lucency of the spinous process of D12, in addition there is thought to be an area of destruction involving the pedicle of D-12." The X-rays had revealed that her lower thoracic spine had been destroyed by a tumor. And it was not just her lower back that appeared to have metastatic lesions. On the lateral skull survey, "large destructive lesions" were noted in the frontal and parietal portion of her brain. Her right femur about halfway down her leg had a "large lytic metastatic lesion," and her pelvis had been invaded, metastatic lesions having formed on her right side, where her femur was attached to her hip, and also at the tip of her illium. The report also noted an "area of destruction involving the 6th left rib."[4] Maude Jacobs's body was filled with metastatic disease from portions of her brain, through her left sixth rib, down to her pelvis and lower right leg.

When Maude went for a surgical consultation, she probably told the surgeon a similar story about the onset and progression of her current difficulties. He noted widespread metastases on her X-ray films, but he was not sure whether the lesions in her pelvis and spine were cancerous. We know from the record that he palpated her right breast and was able to move the large lump relatively freely since he stated that it appeared not to be attached to other structures. Maude felt no pain as he moved the lump this way and that. But, when he probed and pushed into her right armpit, she might have jumped or winced since, because it was sore, his manipulations must have been painful. The surgeon reported that he felt matted lymph nodes in her armpit, and these would likely represent the spread of her cancer—and an ominous prognosis. When he pushed on her lower back, Maude again felt sharp pain. And she must have been relieved when he began to test the reflexes in her extremities since he did not have to work very hard to elicit the responses he was seeking. The surgeon's report evaluated Maude thus: "Hard, irregular approximately 8–10 cm mass upper portion of right breast—non-tender and mobile, apparently not attached to the chest wall but nipple is inverted. Tenderness in R axilla c . . . matted nodes anteriorly."[5] In spite of the reported mobility of the lesion, which might suggest surgery, she was put on a course of chemotherapy. The surgeons may have decided that her tumor was too large and the cancer too far advanced for a surgical option and that a course of chemotherapy was in her best interests. They may also have reevaluated her tumor since Saenger's summary of the patient's history notes that the lesion was attached to "the underlying structures," rather than free and mobile. If this were true, then surgery would have been too difficult. In any event, the turn toward chemotherapy meant that the surgeons had abandoned any hope of a cure.

We can only speculate about the role Maude played in the decisions regarding her treatment and what she understood and took in during those crucial days in July 1964. This is an important issue since Saenger and his team have been criticized for their failure to apprise patients of the nature of the TBI treatment and its investigational character. It is possible that cancer was never openly discussed with Maude, either during her admission and chemotherapy in July or later during her treatment by Saenger's team. During the 1960s, it was not unusual for physicians to avoid mentioning the word *cancer* to their patients. *Cancer* was then, as it is today, a fearful term, and many people could not and did not confront it.[6] Since physicians were still held in high esteem (although that was beginning to change rather rapidly), they could tell patients what to do, what type of therapy they would undergo, even something as aggressive as chemotherapy, without raising the specter of cancer. Maude's physicians may have assumed that she would not want to know that she had cancer and presented her treatments as a therapy or medication that would relieve her pain.

This was not an unreasonable assumption since Maude had lived with a serious level of discomfort for a year and a half before coming to the hospital. By then, the mass in her breast had grown to the size of a grapefruit and had badly distorted her body. Her breast was heavy and misshapen, and her nipple had disappeared. Each day, as she put on her bra in the morning and took it off before going to bed, she must have been aware of that lump, which would not go away and was, if anything, continually getting larger. Perhaps it was possible for Maude to live with it and hide it from herself since she was a widow (her husband had died four years earlier)[7] and she could not share her fears with him, nor did she have to confront him with her growing deformity.

There are also other indications that suggest that Maude was likely in some state of denial. We have three letters she wrote to her aunt when she had returned home after her chemotherapy was completed. She never mentioned cancer, although all the letters are filled with stories about her debilitating pains and the difficulties that she was having in coping with her condition. In the last letter, she complained: "I Don't think the Doctors noes what is wrong with me they didn't help me."[8] This suggests that she was not willing to admit that she had cancer and that it had spread throughout her body, although she was painfully aware of her debility.[9] One of her daughters, Sherry Brabant, discussing Maude's condition at the ACHRE hearings thirty years later, reached a similar conclusion: that her mother "was unsure of how sick she was."[10] Sherry could only conjecture about this point since she had been twelve years old at the time and Maude tried to keep her fears and her deteriorating condition from her three children living at home at the time.

When Maude returned home following chemotherapy in the middle of August, she wrote her aunt: "I am so glad to be home with my kids. . . . I pray everything will fall in place."[11] And later: "Prayer help me God Brought me Through."[12] But the pain continued and seemed, if anything, to be getting worse. In the first letter after her return, she complained: "My arm is So Sore I Just Cant Bend My Elbow Sometimes." By the second letter she despaired: "I don't think I Ever Will walk good again." Maude may not have wanted to know that she had cancer, but she was coming to terms with her own personal dilemma and the likelihood that she would never be the same. She tells how she could barely "creep" around the house, how she held on to the furniture for support as she bravely attempted to do a little work—as she put it—and to take care of her children. In the third letter, there is a brief glimmer of hope as she proudly announces that she has been able to put on dress shoes recently. But reality immediately breaks in as she confesses: "It take me few minutes to get on my feet." Her letters probably remained one of her few outlets. For Maude, her life was her children, and she had to hide her distress from then and not admit to herself that they would be left alone some time soon: "When night Comes I So tired and Sore I have to go to Bed early. The Children wants me to Sit up watch TV with them. . . . I feel So Sorry for them Some times. But By the time the Day done I am Done for my Back hurts me So Bad it Just give away."[13]

Encounter with TBI

In his case report, Saenger stated that the chemotherapy treatment had markedly reduced the tumor in Maude's breast but that the metastatic lesions continued to grow.[14] The treatment was not bringing Maude any relief from her daily anguish. Still, in spite of her pain and exhaustion, she continued to keep house and feed and take care of her children. By November 2, three months after she completed chemotherapy, the pain was too much, and Maude returned to CGH.

We have virtually no hospital records of the events that were to follow. We do not even know whom Maude talked with and what options, if any, were discussed with her before she was given TBI. Yet a few of the records for the TBI study do remain, and we also have indirect evidence that throws some light on what happened over the next month. We can surmise that

Maude spoke with Dr. Ben Friedman, the internist, and Dr. Harold Perry, the radiotherapist. Saenger was explicit in reports and publications that he did not prescribe therapy for any of the patients. He claimed that the decision to use TBI and the amount of dose was the joint decision of the internist and the radiotherapist.[15] These comments were self-serving since, during the public controversy, Saenger tried to distance himself from any direct connection with the prescription or the delivering of the treatments, yet his remarks suggest that there was an uneasy alliance between Friedman and Perry. Friedman ran the day-to-day operations of the TBI program, and Perry treated the patients. Each would have been very conscious of his role and his prerogatives. Friedman was in charge of the experimental part of the study. It would have mattered very much to him which patients were chosen and what doses they were to receive. Perry, on the other hand, was in charge of radiation treatments and would have to deal with the consequences of TBI, especially if those complications were severe. He would never have permitted Friedman to decide on whom he (Perry) was to treat. And he certainly would never have let Friedman write dose prescriptions for his patients.

ACHRE, her family, and others have questioned whether Maude gave written consent and how much she was informed about the nature of the TBI study and the possible outcomes. On the first point, we can be very confident: there was no written consent. To begin with, there is no trace of it in the records that have been recovered. Although this alone does not prove that Maude did not sign a consent form, Saenger himself stated that written consent began ca. 1964–65 and certainly not before.[16] There are also blank consent forms in the archival record that suggest that written procedures began in May 1965 since they carry the date the forms were produced, May 1, 1965, and since those are the earliest-dated forms that have been recovered.[17] Mid-1965 would also coincide with the period during which formal research procedures began at the University of Cincinnati, although they would have predated the NIH requirement by half a year.

But, even if Maude had signed the May 1, 1965, consent form, all we would know about her discussions with Friedman would be generalities. The form would have contained her signature and that of a witness to the following statement:

> The nature and purpose of this therapy, possible alternative methods of treatment, the risks involved, the possibility of complications, and prognosis have been fully explained to me. The special study and research nature of this treatment has been discussed with me and is understood by me.

Consent is given for photographs and publication for the advancement of medical education.[18]

Still, this consent form would not have provided us with the crucial aspects of the encounter between Ben Friedman and Maude Jacobs. It would not have described in any detail what had been said about the therapy, the nature of the experiment, and the possible complications. More important, it would have given no indication of how the issues were presented to Maude—the shades of meaning that were given to the words and the hope or despair that was generated by them.[19]

However, it is likely that, during the meeting, Maude was told little about the experimental nature of the treatment, the possible complications, or even her own prognosis. In a 1971 letter to Saenger, Ben Friedman stated emphatically that, after 1965, he went into great detail about TBI, explaining that it was an experiment, telling the patients that they might not benefit from the treatment, and so on.[20] The letter is interesting on a number of accounts. Since Friedman had every reason to place the start of formal consent discussions with patients as early as possible, we can be fairly certain that they did not begin until 1965. Indeed, since he made no claim of any type about consent discussions prior to 1965, we can assume with some confidence that the discussions with the patients varied considerably throughout that period and that very little was discussed about TBI experiments and the problems the patients might encounter. Moreover, his letter was written in response to a phone conversation in which Saenger presumably tried to find out what Friedman had been telling the patients about the nature of the TBI treatments. In 1971, Saenger was in the midst of public controversy, and the last thing he wanted from Friedman was a letter stressing the experimental side of the work and the possibility that patients might not benefit from TBI. Saenger, of course, was arguing that he was primarily doing therapy and that the only experiment was gathering response data. Friedman either misread Saenger's query or did not want to give Saenger what he sought. Either interpretation leads to the conclusion that Friedman viewed the TBI program as primarily experimental.

Friedman entered the consultation with Maude in November 1964 very much immersed in developing bone marrow transplants, and he would have been keen to continue to have patients enrolled in the TBI program. We also remember that he and Saenger were having difficulty finding suitable patients in whom to escalate the dose of TBI, and he would have tried to make sure that Maude went into the study. Maude was, as we have seen, in a state of denial and desperate for some relief from her painful condition.

Friedman would have quickly realized this, and he would have presented TBI as something that might help her. At this early date, he would not have discussed TBI as primarily an experiment, nor would he have emphasized that she might have depression of her blood cells due to damage to her bone marrow and how that could lead to bleeding, infections, and even death. Maude, on the other hand, would have listened for anything that would convey some hope, and she probably would have tuned out any references to a "study" that Friedman may have briefly mentioned.

Whatever was said during Maude's encounter with Friedman (and, later, Perry), we know that she emerged understanding little, if anything, more about her medical condition. She also clearly knew nothing about her role as a proxy for a soldier. During testimony before an ACHRE hearing in 1994, her daughter Lillian thought it ludicrous to even consider the possibility that Maude was told about an experiment. Lillian claimed that, following TBI, she never left her mother's side and that she had no idea that her mother was part of a study. She also stated that, during public disclosures about Saenger's experiments in the early 1970s, she had argued with others that her mother "had nothing to do with it," that is, with the TBI experiments.[21]

Four days before TBI commenced, Maude had blood drawn for immunological studies to be carried out at Fort Knox and for hematological, chromosome, and other studies at CGH. These tests were repeated twice prior to her treatment, and a urine analysis was also conducted.[22] With these assessments, Maude had been calibrated as a human dosimeter. According to the TBI protocol, the investigators were expected to proceed only if the patient's hemogram was stable,[23] if she was in generally good nutritional state, if she had normal kidney function, and if she had disseminated disease.[24] With these conditions presumably met, Maude went on to TBI on November 7.[25]

Maude Jacobs, no doubt nervous and uncertain about her future, must have arisen early that day since her therapy was scheduled for the morning.[26] She might have thought that at least she would not have to sit around all day waiting for the treatment. When she arrived in radiation therapy, she would have been told to change into a hospital gown and to wait to be taken to the therapy machine. She may have met Perry that day, and he may have said a few more things about the treatment. She was more likely passed on to the radiation technologists who were in the front of the room where she was to be treated. Maude would have been led into the room down a narrow corridor that opened out onto a brightly lit and forbidding space filled with objects whose purposes she could not possibly appreciate. There would be thick heavy blocks of different sizes and shapes scattered about, white masks and body molds covered with strange markings, and assorted metallic and plastic objects. The room would have seemed foreboding since there were neither windows nor natural light, only the oppressive and austere feeling of being in a cell or a dungeon.[27] We do not, of course, know whether Maude had these very responses or whether she noticed any of the devices in the room, but she could not have missed the large machine dominating the room—especially its enormous bulbous head with "El Dorado" written across the front.[28] She would have realized right away that it was the therapy machine, the cobalt unit that would give her the treatment. Beneath it, or, perhaps, swung to its side, she might have taken in something that appeared to be a narrow bench, perhaps for patients to lie on during the treatment. But she was taken, not toward the bench and the cobalt machine, as she might have expected, but to the far side of the room, to a platform with a chair on it near one of the walls. It would have been necessary for the technologists to help Maude up onto the platform because of the pain in her back and her right leg. They would likely have explained to her at this point that this was where she would get her treatment. The chair that she finally sat in was facing one of the walls, not the cobalt machine, which was off to her side. If she had peered over her shoulder, she might have seen it moving in a large arc as one of the technologists stood beside it, watching. As it came to rest, what had been the lower portion of the bulbous head was now facing toward her with its front open like a large yawning mouth.

The technologists would have asked Maude to lean forward in her chair, pulling her legs up toward her like a fetus. They would have said that it was important to hold still during the treatment so that her whole body would remain within the field of radiation. During these moments, one of the technologists would have taped some small objects to her pelvis, her chest, and her head, first on one side, then on the other. If she asked what they were doing, she would have been told that the objects were little measuring devices that were used to record her treatment, or some similar comment that was general enough not to be alarming yet technical enough to identify the devices and explain the process.

Suddenly, the room lights would have been extinguished, except for a bright swath of light that she would have felt shining on her. If she turned for a moment toward her other shoulder, the one farthest from the machine, she would have seen her body projected in shadow onto the wall nearest her, framed in a large square of light, barely enclosing her head and feet. Her framed body on the wall may have reminded her of the shadow games that children play, where they make their hands appear to be a rabbit or a bird in flight. But, before she could have mused for very long, the room lights would have come back on, and the technologists would have started toward the entryway as they instructed her to hold her position until the treatment was over. That would take almost half an hour, and she would also have been informed that they would be watching her and that she could call out if she needed any assistance. After what must have seemed an eternity, the technologists would have come back into the room, and they would have rotated the platform so that she was facing in the opposite direction, with the cobalt machine nearest her other shoulder. She was now halfway through her treatment, and the first part would now be repeated on her other side. We know nothing of what Maude experienced during that long wait. According to the testimony of some patients who have undergone similar treatments, she would have had no particular reactions, and she should have felt nothing at all, no pain, no burning, no other physical sensations. She may have felt a bit woozy during the treatment, especially if she sat very still and did not exercise her muscles to help her blood circulate. Her back and legs would also have hurt all the time as she concentrated on getting though the treatment. She probably prayed or communed with God during that endless wait; certainly he had brought her through worse moments.[29]

The only record of Maude's radiation treatment that has been recovered states that it took place precisely between 9:45 and 10:45 A.M., while the actual exposure should have taken fifty-four minutes and six seconds.[30] We know that the latter figure is accurate because the exposure time can be exactly calculated from the prescribed dose of 150 rads. We also know that the cobalt unit would have been set twice, each time to irradiate her for twenty-seven minutes and three seconds, and that a timer on the console of the machine would have retracted the cobalt source at the end of each time interval. The overall duration of the therapy session could not have been precisely one hour since it would have taken more than six minutes for the technologists to position Maude and enter and leave the room. The recorded interval of one hour does suggest, however, that the treatment progressed without incident and delay; otherwise, a longer overall time should have been recorded.

At the end of the session, as the technologists helped Maude down from the platform, she began to feel a bit woozy. Almost immediately, she was carried off by a severe bout of nausea and vomiting. The case history is quite explicit on this point: "At the termination of treatment, the patient had severe vomiting which continued throughout the next 24 hours in spite of intramuscular compazine."[31] Nothing could stop the vomiting or alleviate the violent paroxysms that seized her over and over again. Even after the worst had passed and the seizures had become attenuated and less frequent, she must have felt severely depleted and exhausted from her bouts of vomiting, and she must have been very frightened.

Maude could not get out from under her newly worsened state. The radiation had made her terribly ill, and it had not brought any relief from the constant pain. Her right leg was, if anything, worse than before. She fell into a state of lethargy and depression. She had no appetite. If she forced down some food, there was every possibility that she would become nauseous again and begin another round of vomiting. Her calorie intake had to be near minimal, and she was probably often dehydrated, with headaches and fever. Maude lasted four days after treatment with TBI before she was readmitted to CGH.[32]

Maude's final days were terrible. Within a week of the TBI, her lungs became filled with infiltrates, one of its lobes had collapsed altogether, and she was near to choking to death. Her heart was galloping out of control, trying to pump more and more oxygen into her system. Her immune system was on the verge of collapse. Her white cell count was falling rapidly, as were her platelets. And she continued to vomit. Her symptoms were those of acute radiation sickness, something a soldier in nuclear war might experience. Antibiotics were started to help her immune system fight off opportunistic infections. Heparin was administered for her heart condition, even though her platelet counts were falling and she might begin to bleed internally.[33] Sherry, one of her children, remembered that her mother "was very depressed" and "out of her head quite a bit of the 25 days she survived."[34] Lill was with her night and day. She moved a coffee pot into Maude's room in the hospital so that she could remain at her side.[35] She would go to Maude's apartment each day to shower and check on the three children.[36] Lill had to balance all this as she looked after her own children. The pressure on Lill had to be enormous as Maude would lapse into periods of hallucination.[37] By the third week following TBI, Maude was still vomiting, but now her stomach contents were filled with dark pellets, an ugly and terrifying omen.[38] Her suffering finally ended. The death certificate reads that Maude Jacobs, a white female housewife, died on December 2, 1964. It records her father, John Henry Eldridge, her mother, Martha Turner, and some details of her burial and is witnessed by Mrs. Lillian Murphy, her daughter. The only other detail on the certificate is the cause of death—"Carcinoma of the Breast."[39]

Other Stories

The statement that Maude's death was due to breast cancer implies that something approaching the natural history of the disease had led to her demise. The effects of radiation on her system and its role in her death were not mentioned at all. Although, in Saenger's reports to the DOD, Maude had been constructed as a surrogate soldier, at CGH she was simply a cancer patient. Except for the hematological findings—which are too detailed and numerous for routine therapy—the records that survive present Maude as a cancer patient, not an experimental subject. Even the radiotherapy record does not formally indicate that she was part of a study. It is simply a report of the date and time of treatment and the amount of radiation administered. On reflection, it is clear that the record is somewhat unusual. It is carefully typed and arranged, with the details of the dosimetry and the calculation of the exposure time on the left and the results of dosimeter measurements indicating the doses received on the right. It is also clear that the chart was typed after the treatment was completed—well after since the readings of the dosimeters that had been taped to her prior to the treatment would not have been available for a few days. Such a carefully typed record would not have been produced for nonstudy patients. Nevertheless, there is one clear indication that Maude was part of an experiment. On the top left of the therapy record there is the handwritten number 45.[40]

As Maude passed from Friedman, to Perry, to the technologists at the cobalt machine, to her home, and then back to the hospital, she gradually and successively passed from the role as surrogate soldier and human dosimeter to that of cancer sufferer and patient. To Friedman, she was someone he wanted to enroll in his study. To Perry, she was predominately a cancer patient who needed treatment for her painful metastases. No doubt she was also a potential study patient whom he had discussed with Friedman. Perry also had to make sure that the treatment followed the protocol, and he had to follow her responses to TBI. But, for him, she had more than partially shed her experimental role. For the technologists at the cobalt-60 machine, Maude was a sick patient who would receive TBI. They understood that she was part of a study, but they would have known little or nothing about the details beyond how they were expected to carry out the treatment. It would have been quite unusual for the technologists to play any further role in the TBI project; they would neither have attended any relevant meetings nor have read or contributed to any related reports or publications. By the time Maude returned to the hospital, all visible traces of her role as experimental object were virtually obliterated for those who came in contact with her, except, perhaps, for the detailed blood studies that continued every few days. When she died, she was a person suffering from cancer whose medical problems resulting from cancer were so enmeshed with those resulting from therapy that no one could or would try to disentangle them. Nor was her iatrogenic disease very different from that of a breast cancer patient who had undergone chemotherapy instead of TBI. Such patients could also have had bouts of lethargy, nausea, vomiting, and opportunistic infections.

If Maude was construed for the DOD as a surrogate soldier and for CGH as a cancer sufferer and patient, she had at least two more roles yet ahead for her after her death: as a patient involved in a cancer study according to Saenger in the early 1970s, and as a victim of the cold war according to her children in the 1990s. In a 1973 paper entitled "Whole Body and Partial Body Radiotherapy of Advanced Cancer," Saenger subsumed Maude into various cohorts of patients that had, he claimed, received TBI as part of a study, not for military purposes, but for treating advanced cancers. Maude was eligible for the study since, according to Saenger, she had "advanced cancer for which cure could not be anticipated." She was one of fifteen breast cancer patients who had undergone TBI and whose survival, Saenger reported, "appears somewhat better than that of the patients treated solely by estrogens and androgens but not quite as good as the group treated with 5-flourouracil."[41] Indeed, the median survival following TBI in this group was, he claimed, 445 days. In other words, half the patients survived well over a year, a result that Saenger felt important enough to report as a contribution to the state of knowledge in cancer care.

Although Maude may seem to be lost in the tables and graphs that Saenger presented, we are in a position to find her in the mass of figures. Saenger wrote: "One can identify 8 cases in which there is a possibility of the therapy contributing to the mortality." Maude was part of that group as one of the two who had "extensive previous chemotherapy." She can be found in table IV, where patient 45 is paired with a survival of 138 days following diagnosis and 25 days following 150 rads of TBI. We can also locate her in one of the survival curves, where she is the fourth circle from the origin, one of the poorest performances represented on the graph. She surfaces again as one of the nineteen patients who died within twenty to sixty days. Saenger compared those nineteen patients with another group of cancer patients whom he did not treat with TBI and who also survived between twenty and sixty days. Using a statistical measure to compare the two groups, he argued that there is no survival difference between them. He concluded: "In other patients described [i.e., Maude and her eighteen companions], the effect of whole and partial body radiation therapy was less important in contributing to death than was the extent of disease in these patients."[42] According to Saenger, Maude had died from her cancer, not from bone marrow syndrome; her death certificate got it right. He would take these claims that Maude was a member of a cancer study who died of her disease to the Bryant Committee hearing in the House of Representatives in 1994, to the ACHRE staff who interviewed him, and to the lawyers on his defense team. And, by the time of the hearings, he had also backed off his statement that TBI might have contributed to the death of eight of the patients.

While Saenger saw Maude as one of a cohort of patients who had died predominantly from cancer, her children came to consider her, not as a cancer sufferer, but as a relatively healthy individual who was a victim of cold war experiments. Although, in a letter to ACHRE, her daughter Sherry acknowledged that Maude was in pain, she presented her difficulties in the context of someone who "cleaned and cooked and cared for me and my two sisters."[43] Maude "could possibly have gone into remission" and may have been saved since "a cure could have been found." Maude had not died of cancer but was "murdered" by Saenger.[44] We find such claims over and over again in testimony during the Bryant and ACHRE hearings. Catherine Hager wrote that her father, who had lung cancer, was "working and leading a normal life" prior to the TBI treatment. He had become someone who was "hand picked and used in Total Body Radiation."[45] To Joe Larkins, it seemed that his father, Willard, went "from a fairly able-bodied middle-aged Father and Husband to a premature death" despite having been someone who "possibly could have lived for several more years."[46]

The children also spoke of the devastation that the deaths brought to them and the rest of their families. "These 'Doctors' left my Mother with no job skills, to raise a grandchild," and she died a "broken women after my Father's premature death," Joe Larkins testified.[47] Sherry Brabant felt that Saenger had condemned her "to hell" since, after Maude's death, she and her three sisters were forced to go to an orphanage and then separated.[48] The possibility of coming to terms with death and the potential for healing over time were obliterated for these children. Only pain, anger, and frustration remained. During and after the ACHRE hearings, old wounds were opened up and discussed in the new language of informed consent and human rights. The children could only wonder in exasperation, if not cold fury, how such experiments on patients were possible. Many of them could see only conspiracy and victimization.

These feelings were abetted by the insensitive behavior of the University of Cincinnati's hospital bureaucracy. When Sherry first requested Maude's hospital records in February 1994, she received a form letter stating that "it has been determined that your family member was not among those reported to the Department of Defense" and that Maude's records would not be retrieved unless it was established otherwise. CGH stated that it regretted the delay in responding to her request but it was sure that Sherry would "appreciate the difficulty due to the large number of requests, the age of the records and our desire to be as accurate as possible."[49] When Sherry persisted, CGH answered three weeks later in a memo addressed to "Medical Record Requestor" and ending in the exact same words that had closed the response to her previous request, that is, with regrets and the certainty that she would appreciate the difficulties involved.[50] No wonder Sherry put a note on the first letter: "Mistake or attempt at cover-up."

Even when Sherry finally received the hospital records, she was angry and frustrated when she saw how little information they contained and how hard they were to decipher. During her appearance at public hearings held by ACHRE in Cincinnati in 1994, she and other family members expressed their rage at the condition and paucity of the records they received. Sherry asked the ACHRE staff how it was possible that so much was missing from her mother's hospital chart. "Things like doctors' notes and nurses' notes, the kind that you would have on the clipboard at the bottom of the bed," were all missing.[51] The photocopies that she had received were in very bad condition. There were many places where they could not be read because blank paper had been photocopied over the underlying writing. This may have occurred when the microfiche copies were made more than twenty-five years ago or during photocopying in 1994. During the ACHRE hearings, the president of the University of Cincinnati tried to assure the families that hospital staff had done the very best they could but that the records were over twenty-five years old and badly deteriorated.[52] Such comments provided little relief to the families.

What the university president did not say, and what he would not say under the circumstances, was that the families really should not expect much more. Hospitals have a lot on their agendas, and the CGH records were not produced or preserved for later investigations by families—and certainly not for lawyers and historians. They were written primarily for other hospital workers as they engaged in the acute care of patients and to meet regulatory requirements. The major difference in the structure and information required on hospital charts since the 1960s has been the result of an increase in regulatory requirements, but these have not significantly changed the difficulties in retrieving and understanding records.[53]

Yet the charts should not be judged entirely by our own special requirements. The lack of signatures and dates, the handwritten notes that are almost impossible to understand, actually work within the hospital environment. Medical workers are adept at searching through hospital records and getting to the information they need. They can rapidly decipher the scribbles and scrawls of their colleagues and extract the one or two nuggets for which they are searching. It could not be otherwise since only a fraction of the information obtained about a patient is recorded explicitly. Behind the cryptic notes, there is a world of medical knowledge that is shared among medical workers and provides them with a framework in which they can extract the medical essence from those cryptic notes.[54] The patients and their children were strangers to that framework and all that goes on in a hospital. Maude, remember, experienced some of the strangeness of being a patient. Her alienation was felt because she had to participate as an outsider in a foreign ritual using strange objects and choreographed according to rules that only the hospital workers knew. The entire process—except for the medical effects it was supposed to have—would have been opaque to her gaze. Her children also experienced some of that strangeness as they tried to reconstruct their mother's medical history from the hospital records.

At the end of the previous chapter, we saw how the psychologists had reduced the voices of the patients to verbal samples. As victims of experimentation and conspiracy, those patients were in danger of losing another part of their humanity. The children of Saenger's patients who turned to this construction of conspiracy have found little comfort. Victimization was a powerful and double-edged metaphor; it gave meaning to the suffering of their parents, but it also may have distanced the children by taking away some of their parents' humanity. For example, Sherry's story of her mother's victimization took on such proportions that, in a letter to ACHRE, she wondered, among other things, whether Maude had been buried in Baltimore Pike Cemetery with her organs removed.[55] We should not forget that victimization was only part of the story and that Maude and the other patients have complex and compelling stories to tell—if we only let them speak to us. If we accept Maude with her limitations and strengths—her difficulty understanding her condition and her courage in confronting it—then we help recover her humanity. Her story also reminds us that anecdotal accounts of the individual life-and-death struggles of patients are as much about clinical trials as about the procedural and statistical accounts of the researchers. It is to the latter world that I now return.

Notes

1. For an account of the patients involved in the Cincinnati trials, see Stephens, *Treatment*.

2. See also ibid.

3. Admission Note to CGH, July 16, 1964, IND 102194-F.

4. X-Ray Report, July 17, 1964, IND 102194-F.

5. Female Surgery Consult, n.d., IND 102194-F.

6. Patterson, *Dread Disease*, 167–70.

7. Brabant Notes to ACHRE, "Points to be made...," n.d., IND 102194-F (hereafter cited as Brabant ACHRE Notes).

8. I am reproducing the spelling and grammar in Maude's letters. There would be no purpose in using "*sic*."

9. Jacobs Third Letter to Aunt, "Just a few lines to answer your letter...," n.d., IND 102194-F.

10. Brabant ACHRE Notes.

11. Jacobs First Letter to Aunt, "Will take time to...," n.d., IND 102194-F.

12. Jacobs Second Letter to Aunt, "Just a few lines to let you no...," n.d., IND 102194-F.

13. Jacobs Third Letter to Aunt, "Just a few lines to answer your letter...," n.d., IND 102194-F.

14. Eugene L. Saenger, "Metabolic Changes in Humans Following Total Body Radiation," *Progress Report*, February 1960–April 1966, Case History for Patient 45, DOD 042994-A;7/16.

15. Saenger et al., "Radiotherapy of Advanced Cancer," 672.

16. Ron Neumann, Gary Stern, and Gil Wittemore, Interview with Dr. Eugene Saenger, Cincinnati, September 15, 1994, 62, ACHRE Interview Project.

17. Eugene L. Saenger, "Metabolic Changes in Humans Following Total Body Radiation," *Progress Report*, February 1960–April 1966, Consent for Special Study and Treatment, Fig. 1, DOD 042994-A;7/16.

18. Ibid.

19. A signed form would, however, have provided some legal support for Saenger.

20. Friedman to Saenger, December 15, 1971, "In regard to your phone call of December 13, 1971," DOD 042994-A;11/16.

21. Testimony of Lillian Pagano, ACHRE Public Hearing, Cincinnati, Ohio, October 21, 1994, 22, http://www.gwu.edu/~nsarchiv/radiation/dir/mstreet/commeet/pm01/pl1tran.txt/(hereafter cited as ACHRE Cincinnati Meeting).

22. Reporting Form for TBI Study, "Blood, Urine," n.d., CORP 091394-B.

23. Actually, Maude's hematological profile was zero, the normal value for a healthy individual.

24. Eugene L. Saenger, "Metabolic Changes in Humans Following Total Body Radiation," *Progress Report*, February 1960–April 1966, 3, DOD 042994-A;7/16.

25. Eugene L. Saenger, "Metabolic Changes in Humans Following Total Body Radiation," *Progress Report*, February 1960–April 1966, Case History for Patient 45, DOD 042994-A;7/16.

26. The description in this section is based on the therapy protocol in ibid., 3–5, 66–70, and my experiences as a radiation physicist.

27. For a discussion of radiation therapy machines, accessories, and treatment rooms, see Johns and Cunningham, *Physics of Radiology*.

28. Kereiakes et al., "Active Bone-Marrow Dose," 652.

29. The TBI protocol is described in Eugene L. Saenger, "Metabolic Changes in Humans Following Total Body Radiation," *Progress Report*, February 1960–April 1966, 4–18, DOD 042994-A;7/16.

30. Record of TBI treatment for Maude Jacobs, November 7, 1964, IND 102194-F.

31. Eugene L. Saenger, "Metabolic Changes in Humans Following Total Body Radiation," *Progress Report*, February 1960–April 1966, Case History for Patient 45, DOD 042994-A;7/16.

32. Ibid.

33. Ibid.

34. Testimony of Sherry Brabant, ACHRE Cincinnati Meeting.

35. Pagano Testimony, ACHRE Cincinnati Meeting.

36. Brabant Testimony, ACHRE Cincinnati Meeting.

37. Stephens, *Treatment*, 118.

38. Eugene L. Saenger, "Metabolic Changes in Humans Following Total Body Radiation," *Progress Report*, February 1960–April 1966, Case History for Patient 45, DOD 042994-A;7/16.

39. Death Certificate for Maude Jacobs, December 2, 1964, IND 102194-F.

40. The note could have been written by one of Saenger's coworkers when the study was in progress or afterward. It might even have been written by someone at the DOD—since that was the provenance of the record.

41. Saenger et al., "Radiotherapy of Advanced Cancer," 671, 675.

42. Ibid., 675, 678, 679.

43. Brabant Testimony, ACHRE Cincinnati Meeting.

44. Brabant ACHRE Notes.

45. Hager to Total Body Radiation Subcommittee, "In January, 1994 I began...," April 7, 1994, DOD 042994-A;13/16.

46. "Testimony of Joe Larkins, House Judiciary Committee, Subcommittee on Administrative Law and Government Relations," April 11, 1994, DOD 042994-A;13/16.

47. Ibid.

48. Brabant ACHRE Notes.

49. Kupferberg to Brabant, "We have your request...," February 24, 1994, IND 102-194-F.

50. Medical Record Services to Medical Record Requestor, "Attached are copies...," March 15, 1994, IND 102194-F.

51. Brabant Testimony, ACHRE Cincinnati Meeting.

52. Testimony of Joseph Stegers, ACHRE Cincinnati Meeting.

53. During the postwar period, administrators increasingly considered hospitals to be business enterprises, not the volunteer institutions they had been viewed as before the war (Stevens, *In Sickness and in Wealth*, 310–16).

54. See Risse, *Mending Bodies*, 155–86.

55. Brabant Testimony, ACHRE Cincinnati Meeting.

6 Peer Review

It is the sense of the Committee that a study directed essentially toward the treatment of widespread metastatic neoplasm by whole body radiation would have approval with the understanding that the infusion of stored marrow constitutes a supportive measure. • Memorandum from Gall to Grulee

A little more than a year after Maude Jacobs died on the wards of Cincinnati General Hospital, the surgeon general of the Public Health Service (PHS) of the U.S. government sent his 1966 landmark announcement to all institutions with PHS grants: "No new, renewal, or continuation research or research training grant in support of clinical research and investigation involving human beings shall be awarded by the Public Health Service unless the grantee has indicated in the application the manner in which the grantee institution will provide prior review."[1] Clearance by another interest group, the National Institutes of Health (NIH), an arm of the PHS, had been added to requirements for the production of clinical research. Where previously the ethical probity of research relied solely on the judgment of medical investigators, the new regime required the approval of research proposals by local peer review committees. The review boards would be housed in the medical centers under the authority of local practitioners, yet they would draw their patronage from the NIH. The review boards were mandated to assure that proposals met the NIH's ethical rules concerning informed consent, patient welfare, and risk-benefit analysis. In return, the NIH would permit such research to compete for its munificent investigational grants. Both parties would gain from the mediating role of review boards. The local boards would shield scientific investigators from direct government intervention and assure the government of the ethical probity of medical proposals. At the same time, if a research proposal also passed the NIH's rigorous scientific review process, then the NIH could assure its benefactor (the U.S. Congress) that it was funding knowledge that was not only beneficial but also safe from the stain of future scandal.[2]

Peer review at the University of Cincinnati took hold following the 1966 PHS announcement. The story of the so-called Faculty Committee on Research (FCR) at the university and its role in judging Saenger's changing research program emphasizes the differing aims of and the conflicts between the committee and the researchers. The FCR's actions consistently suggest that it saw its mission as more than simply fulfilling the NIH's prescriptive ethical rules. It also acted to support institutional needs, ensuring that research continued to thrive while protecting the reputation of the medical center. The NIH's ethical rules were applied instrumentally (and with great difficulty at that) to meet the center's local aims. Moreover, the FCR acted as a mediating body between the center's investigators and government regulators by translating government regulations into a locally understandable language.[3] In addition, from the researcher's perspective, the FCR was but one more interest group that had to be enlisted in order to launch and keep afloat a research program. As a consequence, proposals could and were modified to satisfy the interests of the FCR, as would be the case when an investigator sought to enlist any other interest group. Yet the FCR was fundamentally different in that it operated from an ethical platform and was mandated to control research practices.

In the first section of this chapter, I describe the development of peer review at the University of Cincinnati in both its local and its national context. I follow early attempts at initiating ethical review at the university and the difficulties the administration had with the powerful department chiefs, who did not want to give up their authority over research to the FCR. The FCR was able to become an important political voice at the university once the PHS announced that research proposals required local ethical review. In the second section, I posit a framework for the FCR's more implicit and institutional role and distinguish that role from its contractual obligations to the NIH. In the third and fourth sections, I follow the efforts of the FCR to judge research proposals submitted by Saenger's collaborators (Ben Friedman and later Edward Silberstein) to study total-body irradiation (TBI) combined with bone marrow transplantation. In Friedman's case, the committee found it difficult to apply the PHS regulations requiring it to balance the risks and benefits of TBI against the welfare of the patients. In particular, it struggled with a framework in which scientific and ethical standards were tightly woven together, as they are in risk-benefit analysis. Consequently, the FCR was open to the personal appeal of a cancer expert, and consensus was reached, not so much through a formal evaluation of

Friedman's proposals, but through reliance on the virtues and expertise of physicians. By the time of Silberstein's proposals, the FCR had modified its procedures to better demarcate scientific evaluation from ethical review. In this case, the committee could reach a decision on the basis of narrow technical grounds, which was more in keeping with its growing confidence in its institutional role. Over the same period, the centers of gravity of the proposals submitted by Saenger's team changed. Friedman's mid-1960s proposal addressed bone marrow technology primarily for military applications, while Silberstein's late-1960s efforts moved toward more definite considerations of cancer therapy and randomized trials.

The Start of Ethical Review at the University of Cincinnati

Prior to the surgeon general's notice of 1966, the University of Cincinnati administration had attempted to institute some sort of control over medical research practices in response to public concern with unethical medical experimentation.[4] Keep in mind that the 1962 Thalidomide scandal—which triggered new federal laws requiring the Food and Drug Administration to impose stricter regulations on the pharmaceutical industry, thereby structuring ethical review and, in particular, stipulating that human subjects must be informed that they would be participating in research—would still have been fresh in the public memory. Also, the British Medical Research Council had recently endorsed the Helsinki Declaration, the World Medical Association's code of medical experimentation. The NIH's intramural research program had already adopted guidelines for regulating human research. And there were indications that the extramural program would follow suit. Indeed, by 1964, the NIH began drafting regulations to control research among its grant and contract recipients.[5]

In the midst of this volatile environment, at a Faculty Council meeting in September 1964 the dean of the Medical School (Clifford Grulee) asked the director of the Medical Center (Edward Gall) to formulate institutional policy for human clinical investigations.[6] Grulee was evidently introducing research regulations as early as possible in order to protect the university from future scandals. He also, no doubt, sought to take advantage of the charged atmosphere to rein in his recalcitrant faculty. Grulee, like deans throughout the United States, was losing control over his medical staff since research support was inevitably awarded directly to the researchers, not to the administration. Since the level of support had grown so rapidly over the previous two decades, the independence of the medical research staff had increased in proportion. Although deans were able to skim off overhead to help fund their academic programs, they were increasingly dependent on research monies, and they were becoming hostages to departmental fiefdoms ruled by powerful chairs.[7]

The newly formed Committee on Research met later that month and recommended that "prior approval should be sought" for all experimental investigations on human subjects throughout the medical center. Nevertheless, the committee indicated that prior review was a departmental matter under the control of the department chair. It did, however, recognize that the investigator "may be required" to submit his or her proposal to the Committee on Research, but only at the "discretion of the department director and/or the Dean's Office." The dean could seek to investigate a specific case, but under no conditions would he or the Committee on Research routinely review research proposals. Authority remained firmly in the hands of department chiefs. Moreover, the committee's instructions to the faculty that "no steps be taken . . . to develop a standard form for (volunteer) patient consent in connection with human experimentation" was a further concession to researchers, and it left the judgment of ethical probity in the hands of medical investigators. Of course, the committee warned the investigators that these instructions did not free them from procuring consent, but how that consent was to be obtained would remain a matter of individual prerogative, not institutional control.[8]

The Committee on Research had at this early stage cast itself in a subsidiary role to the powerful department chiefs, while the administration and, in particular, the dean remained in a weakened state. Following what must have been a research blunder that could have damaged the reputation of the medical center, the dean again attempted in June 1965 to bring research under more centralized supervision. He wrote to the department heads that a "recent inquiry brought to mind the desirability of distributing the broad guidelines . . . approved by our Faculty Council at a meeting last fall." All that the dean could do, however, was to gently remind the staff about certain broad recommendations, attach "suggested guidelines" from an association of medical colleges, and point out that the Committee on Research had "reacted favorably" to them.[9]

When the Committee on Research met three months later, it drafted new recommendations that incorporated some of the wording in the guidelines referred to by the dean.[10] A one-page research form was produced that should, the committee stated, "be approved and filed in the Department Director's office." The center of authority still remained in the hands of the department chiefs, and all the committee was able to do was to urge them to document intended research. The form itself consisted of a small number of headings under which the investigators were meant to record such things as the purpose of the study and the method for procuring consent.[11] The minimal space allotted to the answers was evidently an acknowledgment that the faculty's time was too important for them to make much effort. The committee's research document played to an independent staff and testified to continued resistance to prior review. Nevertheless, beneath these very tentative moves, the Committee on Research revealed its anxiety regarding the changing political climate. It reminded the faculty of the pervasive concerns of the public and granting agencies about medical experimentation and alerted them that "more rigorous" policies were being developed at other centers. It expressed its serious concerns about the protection of the medical center's researchers, even of the "status of the Center itself."[12] The overall tone of these early efforts was twofold. First, the aim of the reformers was to try to control the activities of the medical staff, to make them accountable in some fashion to the administration and to their peers. The faculty overall was obdurate in responding to these efforts and sought to maintain their prerogatives. Second, the purpose of the committee's efforts was primarily, if not entirely, aimed at the protection of the reputation of the medical center.

The surgeon general's memorandum of February 1966 abruptly changed the political landscape. It realigned the power relations between the Committee on Research and the medical staff since it required that all research proposals pass through the committee and gain its approval before proceeding to the NIH. In the long term, the policy had institutionwide implications. Not only did all NIH research and training grants immediately come under the purview of the local peer review committee, but, for legal reasons, all research within the medical center would eventually have to subscribe to the same process. In addition, the surgeon general's memorandum reconstituted the role of the committee, requiring it to produce a set of governing principles for its review process. The memorandum suggested that local committees consider developing their program on the basis of the three principles of the rights and welfare of human subjects, informed consent, and risks and benefits.[13] The Committee on Research at the University of Cincinnati, as did similar committees at almost all medical centers throughout the country, adopted in toto the three principles as its standards for local peer review. The committee's role was further strengthened in July of the same year when the NIH rules were modified so that each review board itself certified proposals, rather than each investigator providing the NIH with the supporting ethical documentation. This administrative move was meant to reduce the burden of documentation on the NIH, but it also shifted the locus of responsibility away from the investigator and further elevated the status of the review committee.[14]

The FCR's Ethical Charge and Contractual Obligations

By November 1966, the Committee on Research—under its new name, the Faculty Committee on Research—had, through the patronage of the NIH, become the sole means by which a research proposal could be certified as worthy of consideration for funding by the NIH. Investigators had to submit their proposals to the FCR for certification, and, if they received it, they appended a statement to their grant application that "the investigations . . . in this application have been approved by the committee . . . in accordance with the institution's assurance."[15] Those assurances were ostensibly that the proposal met the NIH's three ethical guidelines.[16] This task, however, was difficult for the FCR to follow for a number of reasons. To begin with, the criteria for informed consent were not specified since the NIH left that to local custom and law. As we saw in chapter 2, the flexibility of the principles was meant to allow them to be used at various localities without coming into conflict with the diverse practices and laws in the United States. The lack of guidelines was, however, especially problematic when reviewers tried to apply the other two principles to evaluate research proposals. The first principle, protecting the rights and welfare of patients, was a clear admonition that placed it in contradiction to the third principle—that of balancing risks and benefits.[17] The use of deontological and utilitarian principles within the same framework would become commonplace in future medical codes of ethics that followed the NIH regulations. How were these contradictory admonitions to be weighed in ethical decision making? According to principlism, the dominant bioethical theory from the late 1970s on, such principles would be applied to each situation by ethical experts who, through rational thought, would resolve the conflicting issues.[18] The situation for peer review committees in the mid-1960s was particularly difficult since they had few, if any, historical precedents to draw on.

Balancing deontological and utilitarian principles was not the only issue at stake. The local review committees also had difficulty applying the risk-benefit principle in its own right. The principle required peer reviewers to weigh present risks against possible gains for the patients and the future benefits to society. This meant that they had to ask questions about the scientific aims and technical procedures of the research proposals in order to assess the likelihood of the aims being met. This assessment of success was important—indeed, crucial—since it had to be balanced against the possible risks. Thus, the risk-benefit assessment forced reviewers to begin judging the scientific details of the proposed research, something that the FCR members had not initially expected when they began to apply the NIH's peer review process. Reviewers were, thus, placed in a problematic and somewhat weakened position, for they had to be able to question investigators more knowledgeable and experienced than they themselves were regarding the issues at hand. This situation was somewhat different from the type of scientific peer review at the NIH (which was the model for local ethical review), which involved the recruitment of scientific experts from around the country to serve on study sections. Locally, such expertise was limited, and review committees had to be able to hold their own against the claims of investigators who possessed specialized and esoteric knowledge.

The disciplining of scientific practices according to ethical determinants was also uncomfortable for reviewers since it ran counter to the predominant (if not universal) view that science works best when unhindered by extrascientific considerations (like ethics). Reviewers preferred to follow a procedure where the "ethical" and the "scientific" could be evaluated independently of one another. One consequence was that, with time, the FCR adjusted its practices toward one in which it could keep the ethical and the scientific content essentially separate, as we will see.

The FCR was able to make these adjustments because it viewed its compact with the NIH more broadly than simply as providing an ethical assessment. Explicitly, the NIH would not consider a research proposal unless a local review committee had certified that the rights and welfare of patients would be protected and balanced against the benefits and risks of the research. Trustworthy knowledge would be produced by research proposals that met these criteria. This explicit compact, however, had an implicit and more institutional character. As I demonstrate below, the local review committees' practices were intended to strike a balance between protecting the reputation of their medical centers and promoting the certification of research proposals. Trustworthy knowledge, in this sense, was, not about patient protection and research benefits, but about crafting proposals that would pass muster scientifically and ethically in order to support the reputation and goals of the medical center. That local review committees would give precedence to this rationale should not be surprising since the NIH framework was meant to place trust in medical institutions over and above the individual actions of researchers. Indeed, as we saw in chapter 2, the NIH framework was an institutional move that was put in place not primarily to protect the welfare of patients but to restore trust in the medical community and maintain the vast research enterprise. The protection of patients and the possible benefits of research were instrumental to institutional goals. And the local review committees understood that the game was about trust in institutional practices.

Although subsequent sections will demonstrate the FCR's focus on institutional matters, we have already observed an overtly institutional agenda in the dean's concerns about a recent inquiry and the worries of the Committee on Research about granting agencies, public exposure, the center's future, and the welfare of its researchers. The following sections show that the contractual arrangement between the NIH and the FCR should be interpreted as an institutional practice, which, in turn, explains why the FCR's interest in Saenger's research project grew and lessened with the changing political climate and why, in spite of significant concerns for the welfare of the patients, the FCR approved Friedman's and, later, Silberstein's proposals.

The FCR and Friedman's Proposal

In March 1966, Ben Friedman and Eugene Saenger submitted a proposal to the Committee on Research entitled "Protection of Humans with Stored Autologous Marrow."[19] Their aim was to study whether bone marrow that had been aspirated from a patient, stored, and then reinfused following TBI could repopulate the marrow spaces depleted by the radiation treatment. Recall that, as early as 1963, Saenger had realized that he could not increase the dose of TBI any further (beyond 150–200 rads) since the patients were suffering from severe malaise, infections, and bleeding owing to the destruction of their bone marrow by the radiation.[20] In the 1966 proposal, Friedman remarked: "Severe hematological depression occurred in all 16 patients who received more than 125 rad total body radiation."[21] And, according to Saenger's contemporary report to the Department of Defense (DOD), that condition "was found in most patients who expired."[22] Maude Jacobs, who received 150 rads, was one of those patients.

Although Friedman had already extracted marrow from thirteen patients, he had attempted to reinfuse it in only two. The investigators had a sense that the transplant may have taken in one of the patients, but their data were ambiguous, and they were reluctant to present this finding as more than a tantalizing possibility. Their quandary was typical of the uncertain state of knowledge regarding bone marrow transplants in the mid-1960s. The literature was, as we have seen, filled with innumerable questions about how much marrow was sufficient for a transplant to take, how it should be stored, what was the best form of filtration (or even whether filtration was necessary), and when it would it be best to infuse it. All these questions had a bearing on the safety of the technique and whether lasting bone marrow growth in recipients was possible. For example, Friedman has been using the method developed by Kurnick, which did not include bone marrow filtering, and there was some concern that this could increase the chance of pulmonary embolisms. Then again, filtration could reduce the possibility of a marrow take: no one really knew.[23]

Friedman proposed to divide the patients into two groups that would each receive 150–200 rads of TBI and an autologous transplant either one day or three weeks after radiation treatment. He proposed to compare peripheral blood counts and bone marrow biopsies for these two cohorts against those of the sixteen patients previously irradiated to this dosage range. Later, he also intended to develop other methods of storage and infusion. Friedman attached to the proposal a consent form that had been developed in 1965, and, in the covering memorandum, he remarked that the form would be signed after the patient had been advised that the nature of the project was investigational and that "therapeutic effects are hoped for, but not assured." He acknowledged that the study carried a hazard of radiation injury, but he argued that "the benefits of possible proof and improvement of marrow storage methods seem to justify this approach."[24]

The FCR's response to Friedman consisted of a running list of criticisms that had been compiled from the responses of the various reviewers. One questioned whether "the radiation is admitted as therapy or purely as an experimental maneuver." He went on to request a more detailed description of the effects of TBI on the previous sixteen patients. Another wondered whether the courses of the patients' diseases were influenced by TBI in ways other than that indicated by the hematological measurements. Yet another inquired whether the patients would be informed that there was no specific benefit, even though the proposal suggests a possible (though poorly defined) therapeutic advantage. The review ended with the bromide that, given the hazard, the committee had "a special obligation to be convinced that the data . . . will be of benefit to 'mankind.'" Underneath the utilitarian conclusion, the committee conveyed a sense of mistrust about the investigators' goals. Its tone was one of disquiet, disbelief, and probing for the facts of the matter.[25]

To some degree, Friedman had himself to blame for this state of affairs. In the covering memorandum accompanying the proposal, he justified the hazards to the patients in the face of the gains in the technology of transplantation, relegating any therapeutic benefits to the background. His work, he claimed, might lead to "an improvement in marrow storage methods," and, even if the methods did not result in viable takes, they would offer a "test system for other methods." The ultimate goal, he announced, was to develop "the best possible way of handling marrow."[26] Even though, earlier in the proposal, he had argued that marrow transplantation might turn out to have a significant therapeutic role in treating advanced cancers, his rhetorical strategy, which focused only on the technical gains, left the reviewers concerned that he had little else in mind.

Friedman's cavalier attitude toward the sensibilities of the committee was consonant with that of the senior investigator (Saenger), who (as a laboratory director) only grudgingly recognized the standing of the committee. In an interview some thirty years later, at a time when he was under intense scrutiny for unethical practices, Saenger could not refrain from remarking that the FCR had no jurisdiction over the proposal and that he had submitted it only as "a good citizen."[27] His claim was that, at that point, only NIH grants required prior approval by the FCR, other medical center research being still under the purview of department and division directors. Although Saenger was, indeed, correct about the standards of the time, his good citizen claim rings false. Friedman's proposal was written on a standard NIH grant application form, and the investigators probably submitted it to the FCR because they intended to send it to the NIH.

Friedman took almost fourteen months to respond to the committee's questions, finally resubmitting his proposal in May 1967 under a new title, one that emphasized its therapeutic goal: "Therapeutic Effect of Total Body Radiation Followed by Infusion of Stored Autologous Marrow in Humans." By that time, he had aspirated marrow from sixteen additional patients and performed transplants on six of them. The committee appears to have taken no notice that TBI with transplantation had continued without its approval in the intervening fourteen months. It may be that it was not concerned since the proposal had not left the institution and, thus, could not damage the center's contract with the NIH. If the FCR's obligation to protect the rights and welfare of the patients was its primary role, rather than being simply instrumental to its institutional agenda, it is hard to imagine that it could have ignored the continuing experimentation with patients.

The primary aim of Friedman's revised proposal was to test whether TBI followed by a bone marrow transplant was "effective palliative therapy for metastatic malignancy in human beings." Its second aim was to investigate techniques of marrow storage and infusion, similar to the earlier proposal. Friedman also made the striking claim that, if a transplant overcame the bone marrow syndrome, TBI would be superior to drugs since "the margin of safety separating marrow death and lethal effects on other systems is smaller with chemotherapy." His argument was that, if bone marrow rescues worked, the amount of tumor cells destroyed with TBI would be higher than with chemotherapy for the same level of normal tissue morbidity. If true, Friedman's hypothesis could have extremely important consequences. He naturally proposed to add chemotherapy and an untreated arm to the proposal to investigate this claim, although, as with the TBI-alone arm, the untreated arm would be culled from retrospective studies.[28]

In spite of its therapeutic claims, Friedman's revised proposal was not successful, in part because his covering letter hardly attempted to answer the FCR's numerous questions, especially the request—left unfilled after fourteen months—for additional data on the clinical effects of TBI. His strategy was to present a more qualitative response along the lines that the current therapy with TBI was toxic enough to justify introducing transplantation even with its additional risks, but not so toxic that continuing the project would be unjustified. Perhaps Friedman was making such a generalized response because he was concerned that the high morbidity the patients had experienced would hurt the proposal. But his unwillingness to directly address the FCR's queries was a crucial error since it left the committee to draw its own conclusions.

What could the FCR deduce from Friedman's proposal? At one point, for example, it indicates that "promordal morbidity [nausea, vomiting, malaise] would approximate 60%." Read as a statement of the consensus of the radiation community about the expected level of complications, this may have been correct, but it hardly answered the FCR's queries about the toxicity that Friedman and Saenger had observed in their own patients. Elsewhere, Friedman declared that "marrow aplasia occurs at 450R," which was another generalized community truth that the FCR had to digest. Perhaps he meant to suggest that the proposed 150–200 rads was far below the standard threshold for bone marrow aplasia, but the committee was left to put these points together on its own.[29] When the proposal finally addressed the TBI patients so far treated, Friedman presented their mean survival times, rather than overall survival rates, which had to raise the suspicions of the committee. Moreover, his arguments were contradictory. On the one hand, he appeared to claim that, since the mean survival times were independent of the dose of TBI, he was justified in increasing the amount of radiation since that would not *decrease* survival times. On the other hand, he also claimed that there was a therapeutic effect with increasing dose, even though that belied the mean survival data.[30] The argument was actually better developed elsewhere by Saenger, but, here, Friedman left the committee (and the reader) to wonder.[31]

The committee was not convinced by the revised proposal. Indeed, if anything, some of the members were more disturbed than before, and they openly expressed their mistrust of the investigators. One of them remarked: "The present proposal has been modified to emphasize, *at least initially*, the potential therapeutic value of irradiation" (emphasis added).[32] A second had the feeling that "the real intent is to work on . . . infusion."[33] A third had "an uneasy suspicion . . . that the revised protocol is a subterfuge to allow the investigators . . . to test the ability of autologous marrow to 'take' in patients."[34] Such mistrust only opened the proposal to an even greater barrage of detailed questions that might have been avoided otherwise.

Friedman, who needed the trust and goodwill of the committee, had compounded his difficulties by mistakenly sending his revised efforts to the dean, rather than the committee. Moreover, in the covering memorandum, he had announced that the "present revision includes adequate clarification of the points in question" and that he "would prefer meeting with the committee" rather than "restating them [the clarifications] in this letter."[35] His refusal to directly answer the committee's questions was bad enough. But his dismissal of the process and his calling a meeting was an open affront to the dean and the FCR since it was their prerogative, not his, to call a meeting. He did not even state why he would prefer a meeting to answering the FCR's questions in writing, leaving it to the committee to draw the obvious conclusion that he did not feel a written response to be worth his time.

The situation had reached a crisis point. Friedman was at his wit's end. He has already written two proposals, and the FCR continued to insinuate itself in every nook and cranny of his work. The committee requested yet more information and expected him to make even more changes, some of which were unrealistic and even contradictory. From the committee's perspective, the situation was no better. The FCR could not extract what it wanted from Friedman, and it was struggling with how to assess the ethics of his proposal. To begin with, it had to decide whether the expected morbidity might exceed some "acceptable" level. One FCR reviewer, who was concerned about the risks to the patients, wanted to reject the proposal outright on the grounds that those risks were too high. "I believe that a 25% mortality rate is too high," he noted. Yet he immediately added. "But this of course is merely an opinion." He felt that he had no guidance for evaluating risks, admitting: "It is difficult, in fact impossible, to balance hazard against potential benefit." Throughout his critique, he was never sure whether the benefits might still outweigh the grave risks, even at a 25 percent mortality level, since, he mused, the investigators might be right—perhaps they *could* prolong life. He also realized that, in order to evaluate the potential gains, he had to interrogate the science, the details of the design, and the aims of the study, but he was reluctant to do since so since "it is not our concern directly."[36] There was a repeated refrain of the committee not entering areas where it did not belong. Another FCR member testified to his reluctance to delve into experimental design. It was "really not the responsibility of the committee," he stated, yet he could find no other alternative since the "good to come is related to the experimental design."[37]

The committee members were genuinely uncomfortable with assessing the proposal's merits in order to balance them against the potential risks. William Curran, a legal scholar who had worked on the ethical and legal regulation of research at the time, was also struck by this implication of the NIH guidelines. He remarked: "Under this guideline [i.e., the NIH risk-benefit assessment], it would seem that the review committee must, in order to review the judgment of the principal investigator on this issue, assess the merits of the research study. There would be no other way to estimate the study's potential medical and scientific benefits."[38] Not only were the newly formed review committees struggling with the problem of ethical review, but those who had been involved in the evolution of standards and who crafted the regulations at the NIH were also genuinely surprised at the implications of the guidelines. It was not the newness of the process that led review committees and scholars to repeatedly comment on the implications. It was the nature of the demand that scientific investigations were subject to extrascientific considerations that made them uncomfortable.

Not only was the committee unable to make any headway with its utilitarian risk-benefit mandate, but it was also unwilling, in Friedman's case, to invoke the first NIH principle to protect the welfare of the patients. In all likelihood, that admonition would have forced it to reject the proposal. The committee preferred, if at all possible, to resolve the conflict and allay its concerns through a request for an amended proposal, rather than outright rejection. Consequently, it called a meeting with Friedman. Its members now confronted the cancer experts face-to-face (Friedman and a senior chemotherapy colleague), a situation in which Friedman's previous assurances that his studies were worthy were able to take on greater weight with the committee. The experts could eschew facts and figures in favor of an appeal to trust the researchers' opinions since they possessed "insider" knowledge. For example, Friedman's claim that, with TBI, the "clinical course has paralleled that of compatible patients treated with other agents" — which the committee had rejected — could, in the confines of a meeting among peers, be accepted as an authoritative assurance. The committee could be assured that TBI was used elsewhere, that it was at least as good as chemotherapy, and that if marrow transplants worked—and there were many indications that they should—there was every chance for improving cancer therapy. The investigators could also assure the committee that they appreciated the risks. Were the complications, they could ask, really that different from those of standard therapy? How could a committee of peers challenge claims like these?

Although there is no verbatim transcription extant of the exchanges that transpired at the meeting, the minutes strongly suggest that consensus was reached along these lines. They contain no new data on the effects of TBI, nor were there any rebuttals by the committee members, or at least none so important that the committee chose to include them in the minutes. The terms by which closure was reached were, however, revealed by follow-up comments by some of the committee members. Either to save face or to protect their own interests, the reviewers demanded explicit assurances from Friedman, assurances that he must have made during the meeting. The committee members simply turned those assurances into their own personal admonitions. One committee member wrote: "I believe a statement should be inserted to the effect the radiation therapy to be used is an accepted form of treatment." Another wondered whether "a simple sentence or paragraph should be inserted to clarify the therapeutic implications and/or possibilities."[39] What these committee members were almost certainly ventriloquizing were the generalized reassurances that had been given by the cancer experts and that had carried the meeting. The proposal went on to receive approval subject to a number of provisos, including a request that it indicate that the "exclusive purpose of the study is to determine therapeutic efficacy of whole body radiation" and that stored marrow was intended only as support.[40] The main purpose of these later additions was to demonstrate that the committee had controlled the practices of its investigators and moved them toward therapeutic goals.

The meeting between the FCR and the investigators had served an important purpose. It provided an environment in which the committee could move away from its confining utilitarian mandate, an area that it found beyond its capabilities and discordant with its scientific ethos. It did not want to delve into the workings of proposals to meet social considerations, nor did it know how to rank risks against social benefits. At the meeting, the FCR members assumed a more congenial and pragmatic role. Working alongside their fellow investigators, they received assurances that the risks were consistent with standards of practice (of chemotherapy and TBI treatments elsewhere) and that the study could potentially improve cancer care. Throughout the process, the investigators had sought and received the trust of the committee.

At the same time, something else happened: the terrain on which TBI stood had changed. When the proposal was first submitted, Friedman focused on the technology of bone marrow transplantation, while Saenger needed transplants to produce more robust patients to further his study of proxy soldiers for the DOD. Now, the primary aim of the study had changed to the therapeutic efficacy of TBI with marrow rescue for advanced cancers. No doubt Saenger and Friedman would still pursue their separate interests, but a different framework supported their studies. The script had changed (at least for the moment) from one about proxy soldiers to one about radiation treatments for patients with metastatic disease. The FCR's role had been, not that of sifting through proposals to stop unethical research, but that of upholding the institution's interests by altering research proposals and crafting them into trustworthy products.

It should come as no surprise that the FCR (as any interest group would have) transformed research. We saw in the discussion of clinical trials in chapter 1 that each enlisted group reinterprets or translates the design of the study and pulls it toward its own interests: the greater the number of coinvestigators, the more translations the study experiences. As the project becomes installed at more and more locales, it becomes increasingly embedded and stable, yet more contradictory and inert. Indeed, multicenter trials were criticized because, as they enlisted a larger number of centers, research questions became less focused and, consequently, produced less beneficial knowledge.[41] Saenger also had to adjust his research to balance the needs of his many coinvestigators—radiation therapists, psychologists, physicists, molecular biologists, immunologists, patients, and so on. In this sense, the role of the FCR was no different from that of any other interest group. In another sense, however, it was different. Although the proposal was certified on one register (contractual and institutional), it was formally approved on another (ethical). The FCR might have assured itself (and the NIH) that the proposal would produce trustworthy knowledge, but the project would later be subjected to public scrutiny to determine whether it had met the prescriptive ethical rules. Its standing as a certified product may have become more, rather than less, precarious.

A New Committee and New Investigators

With the departure of Friedman later that year (1967), the proposal dropped out of sight, yet TBI and bone marrow transplantation continued to main-

tain a high profile. By the end of 1969, Saenger had treated sixty-eight patients with TBI; the last of them, to take one example, received two hundred
rads and a bone marrow transplant.[42] The TBI program also maintained a
conspicuous presence throughout the medical center. First, Saenger, along
with Friedman's replacement, Edward Silberstein, submitted a number of
proposals to the FCR addressing various offshoots of the TBI and bone marrow program. For example, a project entitled "Evaluating Bone Marrow
Granulocyte Reserves in Patients with Metastatic Carcinoma before and
after Whole Body Radiation" was approved in December 1968 even though
the earlier concerns of the committee had still not been addressed.[43] Second, Thomas Gaffney, who had replaced Edward Gall as the, chair of the
FCR, was fully aware of the previous difficulties the TBI proposals had
encountered. He had strongly disapproved of Friedman's proposal on the
ground that he could not "justify 200 rad total body radiation simply for
this purpose [to study marrow takes]."[44] Third, the FCR was, by April 1969,
aware that the NIH had informed the dean that it had previously rejected
"the total body study in patients with malignancy."[45] Finally, at least three
of the members of the FCR were aware that the TBI treatments were still
continuing (even though the project had received only provisional approval subject to provisos) since they were collaborators on the project.
Edward Gall had contributed to the bone marrow phase of the program,
and Evelyn Hess, who went on to replace Gaffney as the chair of the FCR,
and Bernie Aron (a radiotherapist) were included in a TBI proposal that
Saenger submitted to the DOD in 1970.[46]

In spite of the many appearances of the TBI project in the center's activities, it did not (ca. 1970) impinge on the FCR's ethical reviews. The
committee had more immediate worries. To begin with, the NIH had criticized the general consent form for human volunteers that the university had
submitted. Recall that the Committee on Research had earlier expressly refused to develop a general consent statement because of opposition within
the center. Consequently, the university's policy was to use the General
Hospital Admission Consent Form. Naturally, it contained no mention of
research at all.[47] The FCR also faced another troublesome problem when
the NIH held up a proposal (on laser studies) on ethical grounds. The regulators questioned why the committee had approved the proposal. Any
NIH criticism that questioned the judgment of the FCR could have serious
consequences for future submissions and had to be addressed quickly. The
committee responded that the proposal had slipped through its net, not
because of poor judgment, but because "on reexamination . . . the submitted synopsis of the proposed research provided insufficient information

upon which to base a judgment as to gain versus risk." The FCR resolved to require future investigators to submit their entire grant, rather than simply a synopsis of the proposed research.[48]

The FCR was operating in crisis mode. Its main concern was to reassure the NIH that it could shore up its processes and continue to properly certify submitted proposals. It needed to regain the trust of the bureaucrats in the Institutional Relations Branch by providing clear and verifiable responses to their questions. The fact that Saenger's studies were continuing without FCR approval did not engage the committee's interests. Those studies raised no immediate questions about either the functioning of the FCR or the medical center's relation with the NIH. The political invisibility of Saenger's program lasted until July 1970, when, at a meeting of the advisory committee of the university's General Clinical Research Center (GCRC), Edward Silberstein presented an update on the TBI program. A threshold had been crossed. The GCRC's NIH grant would soon expire and TBI, which was part of the last grant, was to be included in the renewal.

Evelyn Hess, who was now chairing the FCR, wrote to Silberstein in August that his presentation at the GCRC meeting alerted her to the substantive changes in the TBI program. Hess wanted a full review of the TBI project. She gently suggested that it might be "wisest" to resubmit the project since it was "originally approved in rather fussy fashion 2 or 3 years ago before you came here" and because the FCR was now operating under new NIH rules.[49] Hess was trying to play down the past problems. She did not even take Silberstein to task for having recently submitted a renewal form even though the project had never been approved. She must have been aware of its status since, in a letter to a consultant reviewer later in November, she noted that the original grant had previously been "given only *Provisional* approval."[50]

The proposal that Silberstein finally submitted raised the ire of many on the committee. A meeting at which the investigators could address the criticisms was quickly scheduled for November 30. In a November 19 letter to Silberstein, Hess instructed him to submit his answers to the committee's questions in writing prior to the meeting. The criticisms were prefaced with a scathing commentary:

> As you may know, the whole study of Therapeutic Effect of Total Body Irradiation was given only *Provisional Approval* by the Faculty Committee on Research in 1967. It is for this reason that I think that it should be completely re-evaluated and a full new proposal submitted. This study has been on-going for a number of years now and we are told that 70 patients have received

irradiation and that the clinical course has paralleled that of comparable patients treated with other agents. The investigators refer to Protocols A and B in this respect. They also refer to Protocol C in regard to the Immune Studies. However, none of these protocols were with the application. The committee requests a full progress report on the data of the 70 patients treated so far.

The specific criticisms that followed went over the same ground as did those of Friedman's submission. The wording barely concealed the reviewers' rage and contempt. The critique ended with a cutting rebuke: "It is noted that the references are all fairly ancient."[51]

Saenger and Silberstein did not show up for the meeting on November 30, nor does it appear that they notified the committee that they could not attend. In a contrite letter written that same day to Hess, Silberstein apologized "for our inability to meet with your committee." The source of some of the FCR's criticisms (e.g., ancient references) becomes clear from Silberstein's confession that the proposal that had been submitted to the FCR was not his own but Friedman's old protocol. He assured Hess that, now that he had read the protocol, he appreciated the questions the committee had raised. He agreed that it was a poor proposal and that he would rewrite it entirely. Silberstein could only "blush that I permitted my name to be substituted for Ben Friedman's." He wrote that he would also try to "dig out of Gene Saenger's old file" two of the missing protocols, but the third one, he chided Hess, "is yours, of course."[52]

Silberstein addressed many of the questions that had been raised in Hess's letter point by point. He informed her that, while he could not provide statistical evidence for the results of TBI in the patients, "Bernie Aron [a member of the FCR] will concur that a striking number of patients . . . have a remarkably benign course." The main thrust of Silberstein's letter, however, was his attempt to forge a new bone marrow program that differed from Friedman's, which, he argued, had not produced "clear evidence of a successful marrow transplant." Silberstein would not store the marrow, proposing instead to infuse it within hours of TBI. Immediate infusion, he claimed, would improve bone marrow survival. Since the marrow was given before the patient's blood counts had a chance to drop, he would, he claimed, be able to identify whether recovery was due to the graft or to the regeneration of the patient's damaged marrow.[53]

Silberstein resubmitted in January 1971.[54] He added to the proposal a randomized trial comparing TBI with bone marrow transplantation against standard chemotherapy (5FU). He also excluded patients whose advanced tumors would have a vigorous response to 5FU. The reviewers, however, were still not convinced. They continued to focus on the survival rates, and they were upset that clinical data had been provided for only twenty-seven of the patients. Silberstein had omitted all the earlier patients (Friedman's), a move that angered and confused some of the reviewers. The committee also questioned whether Silberstein should remove the randomized trial and submit it as a separate proposal. Finally, Hess instructed Silberstein to make substantial changes to the consent statements.[55]

A meeting was called once again. Silberstein once more was unable to attend, but Saenger spoke with the committee. He urged the committee members to support the proposal since TBI was essential and few other investigators were working on similar studies. Saenger's efforts met with little success. A follow-up report on the meeting pointed out that he had been "a little unfamiliar with some of the details." His claim of expertise must have struck the committee as hollow. It reported that it was Saenger's "*impression* the original 200 rad dose had given better results than the 100 rad dose" (emphasis added). Saenger's weak performance impelled him to capitulate to virtually all the FCR's demands. He assured its members that further details would follow from Silberstein.[56]

Silberstein accepted most of the committee's directives, but he was incredulous that the committee and Saenger had agreed to exclude the randomized trial from the proposal. He complained to Hess: "It is difficult to understand your Committee requesting that we exclude the proposed [randomized] study . . . because at the same time the Committee is asking for evaluation of results of therapy."[57] Nevertheless, Silberstein relented. He submitted another version without the randomized trial, although the proposal did contain a comment that the trial was "under consideration." The response of the FCR was worse than before. It could not "decide whether the proposal concerns whole body radiation, bone marrow transplantation or a combination of the two." One of the reviewers argued that, since the submitted study could not determine whether TBI was better or worse than chemotherapy, it should be randomized and that an appropriate method of evaluation developed. The review ended with two questions for Silberstein. How would he assess the effectiveness of bone marrow transplants, and how would he evaluate the effectiveness of TBI?[58]

Hess has lost control of the FCR. The recommendations contradicted each other and the committee's earlier instruction that Silberstein drop the randomized trial. Hess was unable to meld the disparate demands of the reviewers into a coherent set of recommendations. To resolve the impasse, a subcommittee meeting was held one day later. After some discussion, all the objections were reduced to a single technical point. Would the protocol "really allow for valid statistical data"? The author of the question (Harvey Knowles) met with Silberstein three days later. The two reviewed the protocol and resolved their differences, and Silberstein immediately submitted the third revision. Everyone agreed that the proposal was a "much clearer one."[59] In August 1971, four years after its first submission by Friedman, the proposal was approved.[60]

The history of the proposal under Silberstein differed in some respects from its earlier trajectory under Friedman. Previously, Friedman and a colleague could assert their credibility as cancer experts and bring the controversy to (partial) closure. That was not possible in Silberstein's case. Saenger had become too disengaged to assert his expertise at the crucial meeting with the FCR. The proposal had taken on more therapeutic aims, and Saenger was out of his depth. The reviewers did not accept the claims that he made for improved survival at two hundred rads. Once Saenger failed to redirect the committee's agenda, its members had license to redefine the project. Hess was able to stabilize the situation only by having Knowles direct the conflict to a narrow and verifiable technical issue. Once Knowles was satisfied, approval was assured.

There were also differences in the committee's stance toward peer review. The committee members no longer agonized about whether they had the right to review the design of the proposal. Nor did they say much about balancing risks against benefits. Rather, they interrogated the research methods in their own right and they demanded that the consent statements meet the evolving standards. They moved comfortably between the scientific and the ethical, and they had little trouble requesting changes, whether major or minor. They had primarily assigned ethical judgment to an assessment of the consent statement, and they had effectively (although not entirely) demarcated it from scientific review.[61] Closure could be reached on narrow grounds since the review procedure had well-delineated parts.

An institutional process of procedures and rules had, in this case, replaced the individual judgments of the experts. At the same time, except for a few scattered questions about bone marrow transfusions, the committee was noticeably less concerned with the patients' welfare. The focus of the FCR's ethical assessment on the content of the consent statement, no doubt, played some role in reducing internal debates over patient welfare. It may also be that the local understanding of the toxicity of TBI had changed since Saenger had established the de facto standard by treating more than seventy patients. If so, the committee might have been less troubled by the morbidity of the treatments, which had become more commonplace, and somewhat relieved by Silberstein's evidence that bone marrow takes were reducing the toxicity of TBI.

Saenger's program began primarily, if not entirely, as medical experiments for the military and had no therapeutic research aims. Owing in part to the FCR reviews and in part to the high toxicity, the program with time included bone marrow transplants, and the tenor of the proposal changed toward more overtly therapeutic goals. In addition to the interests of his coinvestigators, those of two other groups had forced Saenger to modify the TBI program. The patients, because they could not sustain higher TBI doses, led him to introduce bone marrow transplants. The FCR, for ethical and political reasons, pressured him to modify his program and make it explicitly a therapeutic trial for treating patients with advanced cancers. In the next chapter, new interest groups enter the picture, including the press and the U.S. Congress, as Saenger's program is exposed to public scrutiny.

Notes

....

All unpublished ACHRE sources are from DOD 042994-A;3/16 unless otherwise noted.

1. ACHRE, *Final Report, Supplemental Volume 2*, 473.

2. For further discussion, see chapter 2; and Guston, *Between Politics and Science.*

3. The FCR was acting as a so-called boundary object. Its role in maintaining distance between and facilitating communication among different disciplines and organizations has been widely discussed. See, e.g., Star and Griesemer, "Institution, Ecology"; Jasanoff, "Contested Boundaries"; and Galison, "Trading Zone." For a general discussion of boundary making, see Gieryn, *Cultural Boundaries.* For the importance of language in creating and demarcating cultural authority, see Shapin and Schaffer, *Leviathan and the Airpump.* For the role of cultural authority in modern medicine, see Starr, *Transformation of American Medicine.*

4. For more details, see chapter 2.

5. See Curran, "Governmental Regulation," esp. 430–49; and Faden and Beauchamp, *Informed Consent*, 202–8 and chap. 2.

6. Evelyn Hess, "Historical Review of Total Body Irradiation Project and the Faculty Committee Reviews," *Appendix to the Suskind Report*, December 20, 1971, 3. I am aware that Hess's report was written in the midst of an internal investigation of Saenger's experiments, a period when the medical center and the FCR were themselves under intense scrutiny, and that may have influenced her narrative. I have, therefore, been careful to use only dates of meetings and direct quotations employed in the report to reconstruct earlier events.

7. According to Ludmerer (*American Medical Education*, 259), until 1980 administrative responsibility for medical departments remained primarily with department chairmen.

8. Edward Gall, "The Faculty Council Committee on Research met on Tuesday...," September 22, 1964.

9. Grulee to All Department and Division Heads, "A recent inquiry brought to mind...," June 28, 1965.

10. Evelyn Hess, "Historical Review of Total Body Irradiation Project and the Faculty Committee Reviews," *Appendix to the Susskind Report*, December 20, 1971, 4.

11. Office of the Dean, "In submitting request for approval of a proposed investigation...," n.d.

12. Evelyn Hess, "Historical Review of Total Body Irradiation Project and the Faculty Committee Reviews," *Appendix to the Susskind Report*, December 20, 1971, 3.

13. See chapter 2.

14. Curran, "Governmental Regulation," 439.

15. Gall to Grulee, "The following represents the form approved...," March 28, 1966.

16. Although, formally, the rules were those of the parent PHS, I drop the reference to *PHS* in favor of *NIH,* the NIH being the primary funding agency for medical research.

17. William Curran, a legal scholar writing contemporaneously with the mid-1960s developments at the NIH, noted: "Protection of the rights and welfare of the subject is a universal admonition" ("Governmental Regulation," 440).

18. For a discussion of and references on principlism and the rise of bioethics, see chapter 8.

19. Friedman and Saenger to Clinical Research Committee, "Subject: Protection of Humans with Stored Autologous Marrow," n.d., 1.

20. Eugene L. Saenger, "Metabolic Changes in Humans Following Total Body Radiation," *Progress Report*, November 1961–April 1963, 17, DOD 042994-A;5/16.

21. Friedman and Saenger to Clinical Research Committee, "Subject: Protection of Humans with Stored Autologous Marrow," n.d., 5.

22. Eugene L. Saenger, "Radiation Effects in Man: Manifestations and Therapeutic Efforts," *Progress Report*, May 1968–April 1969, 17, DOD 042994-A;2/16.

23. Remember that Donnal Thomas also tested his bone marrow filtration technique on his patients. See chapter 3.

24. Friedman and Saenger to Clinical Research Committee, "Subject: Protection of Humans with Stored Autologous Marrow," n.d., 1.

25. Gall to Grulee, "This relates to a request for approval...," May 6, 1966.

26. Friedman and Saenger to Clinical Research Committee, "Subject: Protection of Humans with Stored Autologous Marrow," n.d., 1, 3–4.

27. Gary Stern and Jonathan Engel, Interview with Eugene Saenger, October 20, 1994, 9, ACHRE Interview Project.

28. Friedman to FCR, "Subject: Therapeutic Effect of Total...," n.d., 1–3, 7.

29. Ibid., 7.

30. Ibid., 8.

31. Saenger, "Effects," 124.

32. Radford to Gall, "The present proposal has been...," April 29, 1967.

33. Witt to Gall, "It is not clear to me...," May 9, 1967.

34. Gaffney to Gall, "I cannot recommend approval...," April 17, 1967.

35. Friedman to Grulee, "Enclosed with this letter...," March 7, 1967. Friedman should have submitted the proposal to Gall, the chair of the committee, rather than Grulee. He had to submit it again to Gall, and, in a letter of April 13, he stated that the committee should contact him if it had any questions. Friedman to Gall, "Enclosed is a revised protocol...," April 13, 1967.

36. Shields to Gall, "I regret that I must withdraw myself...," March 13, 1967. Gall had passed the proposal on to Shields, who responded (in confidence) prior to the formal submission to the committee. Shields wanted to withdraw (although, ultimately, he did not) "for reasons of close professional and personal contact with the investigators and with some laboratory phases of the project." This might explain why he had access to the survival rates.

37. Knowles to Gall, "This concerns the research proposal...," May 17, 1967.

38. Curran, "Governmental Regulation," 441.

39. Witt to Gall, "In regard to the recent committee deliberation...," May 17, 1967.

40. Gall to Grulee, "The committee on Research has for several months...," May 22, 1967.

41. Gehan and Lemak, *Statistics in Medical Research*, 136.

42. Eugene L. Saenger, "Radiation Effects in Man: Manifestations and Therapeutic Efforts," *Progress Report*, May 1968–April 1969, 1, 4, DOD 042994-A;2/16.

43. Gaffney to Gall, "The Faculty Committee on Research has reviewed...," December 9, 1968.

44. Gaffney to Gall, "I cannot recommend approval...," April 17, 1967.

45. Chairman Faculty Committee on Research to Members, "NIH Review of our Faculty Committee on Research...," April 18, 1969. The NIH never informed the investigators that their proposal had been rejected on ethical grounds. There is no evidence that either the committee or the dean discussed the problem with Saenger. The whole affair appears to have been dropped.

46. Defense Atomic Support Agency (DASA), "Radiation Effects Section Medical Advisory Meeting," 1970, http://search.dis.anl.gov/ (accessed July 2001; hard copy in author's files). During the meeting, Saenger's upcoming (1970–75) DASA renewal was presented. Hess was listed as a consultant, while Aron, Horowitz, Kereiakes, and Silberstein were identified as collaborators.

47. FCR, "Minutes of the Faculty on Committee Research Meeting," September 22, 1967, 2.

48. Ibid., 2–3.

49. Hess to Silberstein, "You may remember that this project...," August 28, 1970.

50. Hess to Mauer, "As chairman of the Faculty Committee on Research...," November 19, 1970.

51. Hess to Silberstein, "Thank you for agreeing to meet...," November 19, 1970.

52. Silberstein was pointing a finger at Saenger for pressing him to submit an old, rejected version. Such behavior would not be surprising for Saenger. The figure that emerges from the archives is one who could act rashly and was disdainful of the committee's prerogatives.

53. Silberstein to Hess, "Thank you for your letter of November 19...," November 30, 1970.

54. Hess to Clark, "We have now received the revised protocol...," January 19, 1971.

55. FCR, "Meeting on Dr. Silberstein's Proposal," February 16, 1971.

56. Hess to West et al., "A Subcommittee of the Faculty Committee...," March 9, 1971.

57. Silberstein to Hess, "Thank you for your letter of March 26...," April 6, 1971.

58. Hess to Silberstein and Saenger, "Recommendation regarding approval or disapproval...," July 22, 1971.

59. FCR, "Following the meeting of the Faculty committee...," n.d.

60. Hess to Silberstein, "Thank you for so quickly...," August 3, 1971, DOD 042994-A;5/16.

61. Ethical and scientific considerations cannot be kept separate, although scientific investigators would like to keep them as far apart as possible. The strongest breakdown of the boundaries between them occurred in therapeutic AIDS trials in the United States beginning in the mid-1980s. Interest groups for the gay community participated in designing research protocols, which, in some cases, led to nonrandomized clinical trials (with poor success, as it turned out). For a discussion of the AIDS trials, see Epstein, *Impure Science*.

7 Public Disclosure

All concepts in which an entire process is semiotically concentrated elude definition; only that which has no history is defineable. • Friedrich Nietzsche, *On the Genealogy of Morals*

The effect of disciplinary failure is to suggest, not that the exercise of discipline is itself mistaken, but rather that there is a need for more knowledge about the person or persons to be acted on. • Barry Hindess, *Discourses of Power*

....

"I was shocked and disturbed to learn from today's *Washington Post*," Senator Edward Kennedy wrote to the secretary of defense, that your agency is "sponsoring research in radiation effects in human beings without informing the individuals involved of the military purposes of the irradiation." The October 8, 1971, story to which Kennedy was referring carried the title "Pentagon Has Contract to Test Radiation Effects on Humans," and it covered the eleven years of study at Cincinnati on "how irradiated troops might function on the battlefield."[1] Kennedy's remarks on the floor of the U.S. Senate marked the opening of a protracted battle between the government and the University of Cincinnati over the character of the total-body irradiation (TBI) studies as well as over whose version of the governance of medical/scientific research was to prevail.

Kennedy spoke to the government's stake in the protection of human rights. He argued that the TBI project represented "an incredible infringement of individual liberty and . . . a dangerous precedent for the reduction of human rights in our society."[2] Mike Gravel, Kennedy's Senate colleague, was concerned with the government's role in protecting risks to patients. He was worried that, while government agencies had delegated the guardianship of patient interests to medical experts, some experts, like Saenger, were abusing their trust. Unless Saenger could, Gravel stated, "absolutely satisfy Congress that there is no deceit or extra suffering or accelerated death involved for the unfortunate and helpless people he uses in his experiments," he would support cutting off his funding.[3]

The stakes for the medical community were, if anything, greater than those for the legislators since the attacks on the Senate floor called into question the viability of the liberal rationality that governed medical/scientific research practices. The medical community had built its formidable position in midcentury America on the "professional valorization of competence."[4] Medical competence or excellence was contingent on the production of scientific knowledge for the public good and the development of competent practitioners. Both these rationalities were supported by the medical community's ability to regulate the ethical practices of its researchers and the competence of its clinical practitioners.

The University of Cincinnati team (Saenger and the administration) appealed to these ideas about the liberal practice of medicine as they sought to defend themselves from charges of unethical practices. They claimed that Saenger's TBI studies were in the interests of his patients and that their treatments were free of political and economic contamination. They also argued that the risks to patients and the competence of the investigations were adequately regulated by internal review at the university. To support their position, the Cincinnati team drew a picture of the TBI program that contained sharply drawn boundaries between, on the one hand, the clinical and research programs and, on the other hand, the entire program and the nonmedical world. Within the clinical arena (the Cincinnati team claimed), physicians operated in the best interests of the patients, their therapies uninfluenced by research considerations. At the same time, the military research program was secondary and maintained, as it were, a one-way relationship with the clinical arena. That is, measurements of the responses of patients to radiation could pass from the clinic to the laboratory, but nothing could flow in the reverse direction: patients in the clinic were protected from research initiatives.

Government intervention raised two rather different issues: the risk to patients and the governance of research. The first issue, which was pursued by Gravel, concerned the nature of the TBI program and whether patients had been placed at risk. Gravel engaged the American College of Radiology (ACR), a professional lobbying organization for American radiologists, to investigate the TBI program, the first in a string of internal and external investigations. This approach proved fruitless since none of the reports produced in the course of these investigations was definitive or had a direct influence on the future of TBI at the University of Cincinnati. The second issue concerned who should govern research and focused on access to the patients. Kennedy, as chairman of the Senate Subcommittee on Health, sought to interview the TBI survivors to learn whether they had been properly informed about the investigational nature of their treatments. The Cincinnati team believed that Kennedy's demand was an infringement on the confidentially of the patient-physician relationship, if not the liberal practice of medicine itself. In the end, the Cincinnati team prevailed in the face of immense pressure from Kennedy by arguing on behalf of the welfare of the patients.

There was a kind of symmetry between Gravel's and Kennedy's efforts. Gravel began with a clinical concern about the risks to the patients and was eventually brought up short against research boundaries, which were affirmed by the ACR, while Kennedy started from research questions about informed consent and found himself facing powerful clinical boundaries defined by the patient-physician relationship. The various ways in which the Cincinnati team erected boundaries—so that at one moment they would enclose clinical concerns and at another moment research interests—revealed the very thing they sought to hide, namely, how deeply intertwined and changing were the military and clinical practices in Cincinnati.

What were the consequences of public disclosures of the Saenger case? If we look at the political arena, we see it filled with various groups and individuals making charges and countercharges that cannot be reasonably addressed. At the same time, we also see it deluged with reports from various medical and government committees, reports from other critics and participants, daily news reports, institutional press releases, position papers from Saenger and the other University of Cincinnati participants, memorandums flying back and forth among the participants, and the comments and reflections of various critics and pundits. The volume of paper that was produced was more than any one individual could possibly read and digest. Yet the documents, the meetings, the press conferences, and the reports were just one aspect of a form of investigative government that has become commonplace in the United States since the 1960s. Although the Saenger case does not compare in size and importance with the Watergate scandal or the Warren Commission, to take two prominent examples, it certainly shares common characteristics in the plethora of materials and in how much of the battle was carried out in a public forum.[5] As in these other more well-known examples, in the Saenger case each protagonist claimed allegiance to transparency and openness while battling to release only as much information as necessary. The Cincinnati team made good use of their obligation to protect patients as grounds for withholding potentially damaging information, but Kennedy's staff in similar fashion hid behind their Senate prerogatives.

Nevertheless, in the process of charge and countercharge, certain aspects of Saenger's program were revealed. First, his peers did not identify his studies as unambiguously outside the standards of practice. Indeed, they considered the TBI project as part of the common community of research, and, thus, they did not feel the need to reject it but sought, instead, to reform its nastier features. Second, the exposure of Saenger's program to public scrutiny, as well as its demise, was primarily the result of contingencies. Public scrutiny certainly made the program vulnerable, but TBI was brought to a halt as a consequence of Saenger's reckless behavior. Finally, the TBI program was anything but a fixed entity. It was a dynamic social institution that was interpreted differently by the various protagonists who came in contact with it: the military planners, the patients, the coinvestigators, the Faculty Committee on Research (FCR), and the various critics that I follow in this chapter. In each instance, the program was modified under the influence of these various interest groups. Throughout this chapter, I point out how many of the protagonists tried to fix the program, to define it and hold it in place long enough so that they could either attack it or repair it.

The Great American Bomb Machine: A Story of Military Medicine

The TBI program was opened to public controversy as a result of Saenger's incautious remarks to an investigative reporter, Roger Rapoport. When Saenger spoke with Rapoport in the spring of 1971, following a joint Oak Ridge/Department of Defense (DOD)[6] meeting, his mind was filled with military issues. His group had been highly visible throughout the two-day conference, having delivered three papers in diverse areas of radiation studies. Saenger, who had chaired the opening session, also delivered a paper in another session on human radiation effects under austere conditions. To appreciate the focus of Saenger's interests during the meeting, we need only look at two of the presentations during his session. One was a film of a human subject undergoing radiation treatment to the brain and then performing a battery of motor and cognitive tests. Another described an instrument for measuring the physiological responses of humans in hostile radiation environments that had been tested on patients suffering from chronic leukemia who had been treated with TBI. Saenger reported on the incidence of and latent period for nausea and vomiting in his patients.[7]

Prior to the conference, in the fall of 1970, responding to a request from Rapoport, Saenger sent him material on the TBI program, including some of his DOD reports.[8] In Rapoport's follow-up call after the Oak Ridge meeting, Saenger was probably injudicious in his remarks since, shortly afterward, he sent a letter to Rapoport that attempted to place those remarks in the "appropriate perspective." The letter began where he had left off in the previous conversation, namely, with the point that his studies of the radiation effects of nuclear warfare were "the most important field of investigation today." He added a second rationale, however, that it was also important to investigate systemic radiation effects so that "patients with cancer can be treated with increased probability of improvement (palliation) or even cure." He also enclosed newspaper clippings of a boy he had treated two years previously who had "recently won a basketball shooting contest." He remarked that he was sure that permission could be obtained to use the photograph or the story, should Rapoport care to, and that he would be glad to make the arrangements.[9] Saenger's offer here is important since he would later argue for the anonymity of patients when news reporters and Senator Edward Kennedy requested access to them.

Rapoport's *The Great American Bomb Machine*, which appeared in 1971, was a diatribe against the nuclear warfare industry. In it, Rapoport placed Saenger's work entirely in the context of the American military machine. Saenger's program appeared in a chapter entitled "Offense" that began with a description of the training of soldiers and civilians in the construction and use nuclear weapons. Such training included an understanding of the effects of radiation on combat troops. Rapoport presented tables listing radiation doses and their effects that were used for assessing the combat effectiveness of troops. He then moved on to the scientific basis of the figures. He began with the bizarre tale of Operation Priscilla, which involved seven hundred pigs, dressed in military uniforms, being placed at various locations near ground zero during a nuclear test to measure the radiation effects on them (as proxies for soldiers). The experiment was a complete fiasco; the army covered it up and decided that it would be better to use human subjects.[10]

Following this tale, Eugene Saenger was finally introduced, not as a clinician treating cancer, but as a member of the prestigious National Council on Radiation Protection, the scientific body responsible for recommending low-level exposure standards for the public. He was quoted as opposing those who would introduce more stringent standards since that would be "a drastic setback for nuclear medicine and the nation." Saenger's TBI program was then sketched in the context of atomic warfare: "The patients' contribution to the data usually begins soon after irradiation. Dr. Saenger's team watches to see if the radiation causes vomiting. If so, the experts must note how long the nausea lasts. This sort of information is very important to D.N.A. [the Defense Nuclear Agency, the successor agency to DASA] because it allows them to figure out how long troops will be 'combat ineffective' after receiving various degrees of radiation." Rapoport then asked: Was this research "saving lives"? After providing a brief response by Saenger, he quoted a radiation "expert" who claimed that the study made no biological sense. He wondered why Saenger had been unable to muster a control group, which other cancer studies routinely used. The implications were clear that Saenger had few patients because his study was worthless.[11]

Saenger was portrayed by Rapoport as a nuclear zealot, an incarnation of Dr. Strangelove, ranting over the glory of nuclear medicine and the study of the radiation effects on human populations, and railing against those who would tighten medical and industrial exposure standards since "what is good for radiation is good for the country."[12] Saenger is also quoted as making the outrageous claim that "the threat from this kind of injury [radiation sickness following nuclear attack] is as great in my opinion as the actuality of problems of cancer and far more study of these conditions is required."[13]

It was more than his immersion in radiation studies and military medicine that led Saenger to make comments like these to an investigative reporter. He was, no doubt, aware of the political climate and the adverse publicity that his remarks would likely prompt. Rapoport's initial approach to Saenger occurred in the midst of the extreme social unrest of the spring of 1970. President Nixon had escalated the Vietnam War by bombing Cambodia, and antiwar protests reached a crescendo as demonstrations erupted at hundreds of campuses throughout the country. In Saenger's home state, four Kent State students were killed by Ohio National Guard gunfire during campus demonstrations. The protests were not only against the war but also against any alliance between the academic and the military establishments. There were demands for an end to military-funded research and ROTC military-training programs on campuses. Of course, concerns over pollution (which included radiation pollution) had taken off with Rachel Carson's 1962 Silent Spring, and the resulting environmental movement had grown into a mass affair following the Earth Day demonstrations of 1970.[14]

His willingness to engage with Rapoport suggests that Saenger was seeking an outlet. He had been frustrated by decreasing Pentagon support, even though he argued that "it is most important for us to obtain additional information on the systemic effects of radiation."[15] We observed these same concerns before in some of his DOD reports. More important, perhaps, Saenger was troubled by a changing polity, one demonstrating shifting priorities and increased disdain for anything military. It appears that he could not contain himself from making defiant and outrageous declarations in the midst of bitter social unrest. But Saenger's hubris would turn out to be costly. Rapoport's material shortly became incorporated into the October 8 *Washington Post* exposé, and it formed the basis of Gravel's first assault on Saenger. It led him and the university through an arduous and humiliating period that would not be brought to closure even three decades later.

It might have been otherwise. We should not forget that similar radiation studies had proceeded at other major medical centers without becoming a public issue until a focused investigation by the Advisory Committee on Human Radiation Experiments (ACHRE) commenced in the mid-1990s.[16] And it is likely that there were other medical experiments (radiation or otherwise) that were not under military contract (some of these, no doubt, nastier than Saenger's) and have never come to light. Some critics believe that the public disclosure of the Cincinnati studies was, in some sense, inevitable because those studies were undeniably egregious. On the contrary, public disclosure was precipitated by the principal investigator's lack of prudence.

The University Responds

One of the immediate consequences of Saenger's behavior was a series of sensational reports in the media. The first, in the *Washington Post*, which was based on Rapoport's work, portrayed Cincinnati researchers doing Pentagon experiments on indigent cancer patients who were not told about the complications of TBI and that they were being used for military purposes. The article also reported that the group had published little and that the program was secretive and closed. The *Post* also quoted Edward Silberstein's claim that they were advancing cancer cures and Saenger's pronouncement that military research was "damned important."[17] The latter quote surely could not have helped the medical center's cause. The story spread rapidly throughout the United States and abroad. Within days, the major American news networks (CBS and NBC), news services (UPI and AP), and newsweeklies (*Time* and *Newsweek*) as well as a host of newspapers, including the *Washington Post* and the *New York Times*, sent reporters to Cincinnati and requested interviews, pictures, and special meetings. *Stern Magazine* in Hamburg wanted photographs of the leading figures sent by special delivery. The *Times* of London carried the headline "Hospital Wards Are Not Battlefields." The *San Francisco Examiner* announced "Dying Patients Got Radiation for the Pentagon," while *L'Express* in Paris led with "Les Cobayes de Cincinnati."[18] There were inquiries from Senator Taft and Representative Keating (both of Ohio), and the BBC sent a team to interview Saenger.[19] Within days of the initial story, an emergency meeting was held at the university, at which Clifford Grulee, the dean of the Medical School, Edward Gall, the director of the Medical Center, and Saenger, among others, prepared a press release for a news conference the following day.[20]

The story that was presented at the press conference did not accurately represent the scientific and medical practices at the university; rather, it laid out an idealized version of how medical research should be practiced. This liberal vision of medical/scientific research was a combination of Robert Merton's 1940s characterization that scientific research was carried out by disinterested investigators and the 1966 research regulations of the National Institutes of Health (NIH).[21] The university presented the Saenger team as investigators who readily shared their scientific results with others while selflessly pursuing research guided by the extant regulations designed to protect the welfare of patients with informed consent and peer review.

The university's story went as follows: Saenger began his program in 1955 when he used TBI for palliating advanced cancers. Coincidental to, but *following*, the initiation of TBI treatments, the army learned about the project and expressed an interest in obtaining some of the data. The DOD had no influence whatever on the method of treatment. All military funds were used for laboratory tests only. None went to patient care. The medical faculty had approved the study on a number of occasions following national standards. The patients were fully informed; indeed, the two-day consent process exceeded the standard of practice. TBI therapy was at least as good as, and in many cases superior to, other techniques. The patients were not ruthlessly exploited; rather, the individual medical care and psychological support that they received far exceeded common practice. Finally, the project was anything but secret. All the work was available to the scientific community.

To maintain this liberal vision, Saenger and the administration used a number of divisive tactics to throw up boundaries between therapeutics and research as well as between the medical center and the lay community. One of the most crucial of these was the walling off of the DOD from the clinical program. This boundary was crucial since, on the one hand, it made Saenger's research program independent of crass military ends and, on the other hand, it left TBI as disinterested research and humane treatment. The university's construction hinged to a great extent on a start date. If the project was initiated prior to 1960, when the DOD commenced funding, then the military links could be more readily severed. The importance of determining when the project began is apparent from the great difficulty the Cincinnati team had in assigning a specific inauguration date. Gall's press release of October 11 read: "In 1955 when the present study was initiated...."[22] In an earlier version, the less precise phrase "in the 1950s" was struck out and replaced by the slightly more specific "since 1955."[23] At the press conference, Gall stated something else again, namely, that the study began "in the 1950s, to be exact, it was 1955."[24] In a preparatory note, Saenger was more nuanced when he wrote: "From 1955–1960 ... a protocol was developed."[25] The notion that the research proposal evolved over a period of time is somewhat closer to what we expect from science in the making and is in marked contrast to the public position of the Cincinnati team. At the press conference, however, Saenger did not mention anything like the development of a protocol, simply stating that the study "was initiated back in the 1950s sometime."[26]

With a TBI start date sometime in the 1950s, Gall was able to develop a causal story at the press conference: "Early in the study the investigator acting as a consultant to the Department of the Army described his preliminary results to his colleagues. Interest was expressed in the study since there were implications in respect to well individuals exposed to whole body radiation under other circumstances."[27] According to Gall, the program was already under way when the army approached Saenger, who was more or less doing a public service by sharing his data. Saenger's preparatory notes are again closer to the give-and-take of developing research, and they suggest the influence the DOD had on the project. "The Department of Defense because of its interest in the results of certain laboratory studies," Saenger wrote, "agreed to support certain aspects of this work after reviewing preliminary proposals."[28] The private message, unlike the public one, was that the final agreement followed a research and funding agenda mutually developed between Saenger and the DOD.

In spite of Saenger's more realistic construction of the events, the Cincinnati team was adamant in public that the DOD approached Saenger after the start of TBI and, consequently, that the treatment strategies were entirely free of DOD influence. Gall repeated this mantra during the press conference when he referred to the DOD project as "simply a spin-off." Grulee added that DOD funds only supported laboratory tests and personnel, which was meant to suggest that, since the military was not paying for the treatment, it consequently had no influence on it.[29] The press never questioned this line of argument, though it was not much of a defense. The National Cancer Institute (NCI), for example, rarely supports the cost of treatment, yet it would be nonsense to argue that its interests do not influence treatment protocols on clinical trials.

Other themes characteristic of the liberal ideal of medical/scientific research emerged at the press conference. For example, Saenger and his team were objective and unbiased investigators. There had been no racial or class bias even though all the patients were indigent and a majority were African American. On the contrary, the patients had been chosen from the General Hospital solely on the basis of their disease status; according to Gall, the mix of patients "reflected the type of patient in the General." Moreover, the entire project was open and transparent. Gall remarked that scientific papers had been published in "reputable scientific journals," and Saenger added that "all our data continues to be available," if not to all, then at least "to responsible investigators." Even the charge that Saenger's group had never produced a clinical publication, which, for some, proved that the study was primarily for military ends, was refuted through the investigators' allegiance to liberal science. The team had not published a clinical paper, it was argued, because their data had not yet reached statistical significance. They were, not secretive, but cautious scientists who did not publish immature results.[30]

The starkness of the presentation revealed, not only the team's normative beliefs, but also, perhaps, their fear that the press would exploit anything more nuanced. Indeed, the reporters harped on the university's relationship with the Pentagon and tried to probe into every nook and cranny of any military arrangement, at a cost of not pressing other substantive issues. They asked volatile questions about "whether we have guinea pigs in Cincinnati," and they tried to trap the team with comments like: "Laymen might fail to reconcile . . . that you're supposedly doing the patient good with this kind of treatment yet supplying the Pentagon on its harmful effects."[31]

The encounter was fruitless. It produced more noise than knowledge, and both sides avoided addressing the nature of the clinical program. Since the Cincinnati team claimed that TBI was not being used for military aims, then what was the purpose of the TBI program? Was Saenger doing a *clinical investigation* to extend cures and reduce symptoms, or was he primarily doing *therapeutics* according to the standards of the community? The press appears to have accepted the optimistic treatment outcomes that Saenger presented and did not push the issue. I am not suggesting that some definitive answer would have emerged had the press pursued questions along these lines. Indeed, the position that I have been urging is that the research and clinical programs were both deeply entwined and changing, with the result that a single characterization of the program for the period 1960–70 would grossly distort what was a highly fluid and complex project. Indeed, in the next chapter, we will see that, in the mid-1990s, various groups again revisited the Saenger case and also had great difficulty in defining Saenger's fluid research enterprise. What is evident about the 1971 investigation is that the press was so blinded by the military angle that it let the Cincinnati team off the hook, allowing the researchers and the administration to frame the program as vaguely as they chose.

Although, throughout the press conference, the Cincinnati team drew sharp boundaries between military research and therapy practices, they left vague and uncertain what exactly it was that constituted those practices. By keeping the nature of the investigations as hazy as possible, they afforded themselves the leeway to characterize TBI as the circumstances warranted. Indeed, Gall's rationale for implementing TBI was a masterpiece of obfuscation. He began with the claim that "it was universally appreciated" in the mid-1950s "that there was not successful treatment for advanced widespread cancer." He then moved to the TBI study and argued that "it seemed rational to utilize whole or partial body cobalt-60 radiation for this purpose." But he was deliberately obscure about the meaning of *purpose*.[32] He did not say that TBI was meant as therapy, and he did not follow a background document of Saenger's (which he had, no doubt, read) that claimed: "The purpose of the investigation has been to improve the radiation treatment of the patient with advanced cancer." Gall's public position admitted no human agency, eschewed any mention of the word *investigation*, and did not countenance any definition of *purpose*, even an imprecise one like "improve radiation treatment."[33] Gall was unwilling to enter into any discussion of the character of the TBI program. He wanted the university to hold its fire as long as possible. By the end of the month, he instructed the university's public information office to try to "delay any further features on whole-body radiation until we know what the national climate will be."[34] The Cincinnati team had been definite that military goals were peripheral to TBI, but they were indefinite about the program's purpose. That issue would be left for others to try to unravel.

Three Reports and Further Constructions of the TBI Program

From the late fall of 1970 through January 1971, the TBI program was investigated by the ACR,[35] the University of Cincinnati Junior Faculty Association (JFA),[36] and the Suskind Committee,[37] a group of senior university faculty appointed by the dean. Their political agendas were quite different, and their reports offered various depictions of the program. As a professional association, the ACR produced an extremely favorable report almost parroting Saenger's claims. The JFA, responding to what it viewed as an arrogant abuse of power, wrote a harsh critique charging Saenger with killing patients. The Suskind Committee, as an arm of the administration, sought to protect the medical center and mildly censured, but ultimately supported, the TBI program. Not only were the political agendas different, but the style of the reports also varied according to the relationship each group had to the medical community. The ACR report was an aloof statement by experts, characterized by generalities and claims of insider knowledge. The JFA document was the antithesis, detailed, specific, and armed with quantitative analyses. The Suskind report—a generalized presentation of the events—contained features of both. The different styles were consistent with the argument that weak institutions often turn to quantitative arguments when faced with hostile questioning.[38] The ACR, as a powerful medical group, was comfortable turning to the considered judgments of its experts, while the JFA, as outsiders without any expertise in medicine, could not hope to be taken seriously without quantitative support. Finally, since all the reports relied primarily on Saenger's writings to construct their cases, the divergent readings might be taken as the product of the various political agendas. I do not subscribe entirely to this interpretation. It was rather, as I discuss later, that the varied and changing character of the TBI program made it possible for each group to read the program in its own way.

The ACR investigation was initiated following Senator Gravel's many attempts to enlist medical experts to investigate Cincinnati.[39] The ACR, concerned as it was with the possibility of increased government oversight, agreed in mid-November to look into the matter since, as it stated, Gravel's request might have "an impact on the society." Indeed, it had already discussed the request with Saenger, who was "responsive to have College representatives discuss his project with him."[40] A committee of three eminent physicians from the ACR first met with Saenger at the annual radiological meeting in late November and then spent a day in mid-December interviewing the primary actors in Cincinnati.[41] The report (actually a long letter written by Robert McConnell, the president of the ACR) was submitted to Gravel in early January. It sought to accomplish three things: to reaffirm the distance separating the medical community from the public, to erect a boundary between Saenger's TBI therapy and the DOD, and to support his program and free it of claims that he had harmed patients.

McConnell's letter began by commenting on the distance between the medical and the lay communities. "Physicians," he wrote, "do not invariably share with the public the ways in which they reach professional conclusions." In this case, however, they will do so (and here his condescension reaches its peak) since, he continued, "senators have need of expert impartial medical and scientific advice in evaluating complex biological problems." Not only were the sterling credentials of the three experts paraded before the reader. McConnell also elevated the investigation to the highest level of liberal practices by noting that the committee members "apprised themselves of the situation" to the same extent as members of an "NIH study section or site review team." With these standards of excellence, McConnell felt justified in opening his letter with "broad general conclusions." The project, he affirmed, was valid in its conception, execution, and selection of patients as well as its procedures for informed consent. His final conclusion (in the face of Gravel's threats) was to urge the Senate to support the continuation of the Cincinnati program.[42]

The report also sought to erect a boundary between therapy and the DOD. At the outset, it stated that the committee "viewed the project as it was designed—as a clinical investigation." Yet, like the Cincinnati team at the press conference, it was vague about whether *investigation* referred to some sort of clinical palliation study, or the provision of metabolic data to the DOD as an offshoot to standard therapy, or something else. Nevertheless, it was quite definite that an "investigation" had been planned and that the treatment of patients had not been influenced by military concerns. McConnell mentioned, in a number of places, that DOD funds were used only to support psychological studies and laboratory procedures and that "the DOD exercised no control over patient selection or clinical treatment."[43] He also gingerly addressed the concern that patients who died within approximately the first two months might have succumbed to radiation sickness. Saenger himself had discussed the issue in some of his writings during the late fall of 1970 and had identified eight patients whose depressed bone marrow may have contributed to their deaths. McConnell essentially argued that it was difficult to determine the cause of the patient deaths and that further work needed to be done. Here, he was attempting to turn a defect into a claim for doing more research.

There is little doubt that Gravel was frustrated with the glowing ACR report, filled as it was with generalities. In a February 1972 response, he unleashed a seemingly endless barrage of detailed questions, pointing up the opacity of the report. To take but one example, in response to the ACR statement that survival showed "an extension of days over untreated patients," Gravel asked: "What does the term 'days' mean? 2 days 180 days?"[44] By the time McConnell finally responded to Gravel in March, Kennedy's investigation had been effectively thwarted, and Gravel's threats that funding would be cut off had long proved empty. McConnell ignored Gravel's questions and reaffirmed that the ACR had provided "the considered assessment of three of the most highly qualified cancer experts." There was no more that could be done since, as McConnell ended his letter, we "simply do not think we can improve on the quality of advice."[45] Indeed, the ACR could no more answer Gravel's questions than Gravel could accept the opinion of its experts. Although the ACR report was taken by Saenger's supporters as a vindication, it had little overall impact. Saenger's relationship with the ACR and McConnell and the report's overly sympathetic conclusions undermined its credibility.[46]

While the ACR was sought after as a body of experts, the JFA carried no such imprimatur, but its investigation was driven by the passion of Martha Stephens, an assistant professor of English at the university. When she learned about the controversy surrounding Saenger's studies, Stephens began to badger Gall to let her look at the DOD reports. At first she received only "courteous" refusals. But then, as she later (in 1994) recalled, she received a surprise: "One day I went back over to his office and there was a large pile of documents [the DOD reports] on his desk. . . . Even now, I do not know why Gall surrendered these papers to me." Stephens spent the Christmas holiday break poring over the documents and produced a report for the JFA.[47] Where the ACR report was long, diffuse, and general, the JFA report was succinct and specific. It addressed only three issues: (1) the objective of the experiments; (2) the risks to patients, and (3) informed consent.

I deal here mainly with Stephens's discussion of risks since her analysis is in especially sharp contrast to that of the ACR. Stephens demonstrated that the proportion of patient deaths rose with increasing dose, and she suggested that the data showed a threshold effect in that seven of eighteen patients died when the dose was 150 rads or higher while only two of twenty-two died when the dose was under 150 rads. With such figures supporting her, she then questioned why the doctors continued to use the higher doses after they began to lose patients. She also took the ACR to task for its claim that only eight (10 percent) of the patient deaths (within twenty to sixty days) might have been caused by radiation since her data showed that "14 total-body subjects . . . died within this period (not to mention 5 partial-body) — or 23%, and of course this figure takes no account of the 7 subjects who died within the first 20 days."[48]

As an outsider (and nonexpert), Stephens worried that the medical community might challenge the report. She recalled (again in 1994): "My friends in the JFA had gone over with me every sentence of it. We knew it had to be clear and it had to be right. I introduced a number of qualifying phrases, everywhere we had the least idea that an assertion could be challenged."[49] The JFA report indeed has a sober character that belies in many ways the intense anger that Stephens had (and still has) toward Saenger. Her circumspection was, nevertheless, evident throughout its pages. For example, she remarked vis-à-vis the effect that DOD funding had on treatments: "Throughout the reports to the DOD the doctors make statements that indicate that the selection of patients and the radiation dose given them was at least *partially tailored* to the needs of the DOD project" (emphasis added).[50] The report, like McConnell's letter, had little influence on the course of events in Cincinnati. For the most part, the university administration ignored it and, when forced to comment on it, dismissed the JFA investigation as that of a "small splinter group."[51]

The institution's own investigation, summarized in the Suskind report (so-called after the committee's chairman), aimed, neither to condemn nor to exalt Saenger, but to protect the institution. Its style of presentation lay somewhere between the generalizations of the ACR and the definitiveness of the JFA. It gave many more details than did the ACR report, and, although it mentioned the early deaths, it highlighted, as the JFA report did not, overall TBI survival rates against published data. Suskind focused on the credibility of the institution (through Saenger's survival comparisons) rather than on the welfare of the patients (through how many had died as a result of TBI). The committee was able to reach certain conclusions on these issues since Saenger continually fed them various reports and analyses.

The Suskind Committee was initiated on November 12, a little more than a month after the *Washington Post* article, just two days after Gravel's letter to the ACR, and in the midst of Kennedy's threats to hold public hearings and interview the patients. Dean Grulee urged it to begin its deliberations immediately so that the "conclusions would be available to guide the University in the event that some type of investigation of the studies were to take place."[52] The report (again on instructions from the dean) should make it possible "for the President of the university to make public statements from it."[53] The deliberations, as well as the identities of its eleven faculty members serving on the committee, were to be kept confidential.[54] The report, submitted to the dean in mid-January, was voluminous: it consisted of seven major sections and multiple appendices. (Grulee would brag that it was "three inches thick.")[55] It no doubt made heavy reading then (as it does now), but the crux of it is located, not in the main body, but in the brief summaries and recommendations that follow each section. Indeed, the body of the report does not always support the recommendations, and, in places, the recommendations contradict the report. This is not at all surprising once it is realized that some of the sections were taken (sometimes verbatim) from the writings of the individuals under investigation.

During the entire period of the committee's deliberations, Saenger and his coworkers were producing innumerable analyses of their work. The committee was inundated with this material, much of it found its way into the main body of the report, and it was used to support (and it certainly influenced) some of the conclusions. For example, Saenger had developed survival statistics for different groups of his patients; the most favorable result, for colon cancer, showed some advantage for TBI compared to either chemotherapy or no treatment. His point was that he was doing a clinical study and that his results were no worse, and sometimes better, than standard therapeutics. The writings of other actors found their way into the report as well. The psychology section appears to have been lifted directly from the writings of the investigators, while Evelyn Hess, who had been involved in the FCR reviews of Saenger's program, wrote its history.[56]

The conclusions of the Suskind report reveal the lack of consensus and the evident tension between the impulse to judge Saenger's studies and the political remit of the committee. The report does not distill the varied and complex characteristics of Saenger's program, nor does it really analyze them. Indeed, it is more like a map covering the various byways of the territory. For example, Saenger was apparently vindicated since the committee found that there was "no evidence that the DASA [i.e., the DOD] funding was made contingent on work, ideas, or suggestions proposed by DASA" and that the work "was carried out with complete scientific freedom appropriate for research conducted in University facilities." Yet it claimed that, while the design of the study was adequate to evaluate toxicity, it "lacked carefully selected measures to evaluate palliation." Reading between the lines (if, in fact, that is even necessary), this rather damning statement suggests that the TBI study had been designed to answer military (i.e., toxicity) rather than therapeutic (i.e., palliation) questions. If so, then the earlier conclusion that the DOD had no influence on the study was contradicted. Yet the committee recommended that Saenger's study should be, not rejected, but expanded into a randomized trial comparing TBI with another modality. Note the conclusion of bad practices being turned into a justification for expanding the study, as with the McConnell letter. At the same time, the committee recommended that, given the size of the proposed randomized study, financial support should be sought "from a federal health agency or foundation interested in cancer research," that is, not from the DOD.[57] Saenger recognized the implication of that statement in that it could "suggest that work [his study] has been extremely inappropriate."[58] The Suskind report, like the other reports, would have little direct influence. But the variety of its conclusions provided the president of the university with support for whatever decision he might make.

The contradictory portrayals of Saenger's research were no doubt influenced by different political agendas. McConnell was supporting a colleague and protecting the radiological community. Stephens, troubled by a culture that took advantage of its underclasses, read Saenger's program in the context of the military-academic complex. While disciplining the program to a certain degree, the Suskind Committee mainly supported it in order to protect the medical center. Still, the contradictory portrayals were not simply the result of each group imposing its reading on an unchanging artifact. Each was trying in its own way to mend a social entity that had become fragmented and troubling. And each sought to stabilize the dynamic and socially embedded TBI program and fix it, whether through censure, support, or repair.

In earlier chapters, we saw how the TBI program was anything but a fixed entity. It was a social institution taken up differently by each person who came in contact with it. When Saenger presented his work to his colleagues at DOD meetings, it was about radiation effects under austere conditions. It was those military aspects that Rapoport presented in his book. Friedman and, later, Silberstein had a different view and pushed the program in the direction of bone marrow transplantation. And the patients' response to treatment, as well as the FCR reviews, also, in their turn, led to substantial changes. When Horowitz and, later, Aron brought TBI into the radiotherapy clinic, it was a very different enterprise than what Hess took into her immunology laboratory. What had begun in the shadow of the nuclear battlefield in 1959 had changed dramatically by the early 1970s. And it metamorphosed yet again under the pressures of the fall 1970 investigations. For example, the reports that Saenger produced during the investigations—which emphasized the therapeutic role of the TBI— worked their way into the ACR and Suskind reports as well as into research proposals that Silberstein submitted in the spring of 1972 (more about this in the last section). These later transformations of TBI—when it took on a predominantly therapeutic profile under external scrutiny—should not be considered as fundamentally different from earlier transformations when, for example, it was under FCR review or when it became part of radiotherapy practices. The program was fluid and always appealing to and responding to interests. By interrogating the TBI program, the investigating committees of the early 1970s were, in part, causing it to change yet again.

It was read in so many ways because each critic was, not simply imposing a personal agenda, but interpreting different faces of a social entity. During all these reviews, Saenger was also writing various narratives trying to fix and repair his program. He too wanted to make it hold still long enough to mend it and make it acceptable to his various critics.

Access to the Patients: A Battle over the Governance of Medicine

Nowhere was the question about how medical/scientific practices were to be governed more evident than during the battle over access to the patients that ensued in the fall of 1971 and continued through the spring of 1972. The struggle was initiated by Senator Edward Kennedy, who had been troubled about the Saenger case from the moment the *Washington Post* had reported on it. Kennedy, as chairman of the Senate Subcommittee on Health and a strong defender of human rights, wanted to interview the patients to learn whether there was appropriate informed consent. Saenger and the administration opposed this proposed incursion into their territory on two grounds. First, Kennedy's staff did not have the background and expertise to interview the patients. This is a predictable position, one that most scientific investigators will take when outsiders seek access to the workings of their laboratories. The Baltimore affair of the mid-1980s exemplifies the position that only experts (as defined by the investigators) are capable of understanding and judging scientific work. A young coworker in the laboratory of the Nobel laureate David Baltimore accused one of the senior investigators of doctoring data in a publication.[59] The affair spread rapidly, becoming a national scandal, and leading to, among other things, the NIH instituting an independent office of investigative affairs.[60] Of significance is Baltimore's response to individuals at the NIH who wanted access to his laboratory data in order "to compare them with published data." Baltimore turned down the request because the investigators were "not immunologists and . . . clearly showed a lack of understanding of the complex serology involved."[61] Thus, only experts should have access to laboratory data or, in the case of the Cincinnati case, to the patients.

The second position the Cincinnati team took was, in effect, two related arguments that turned on medical issues. If the medical center revealed the names of the patients to Kennedy, it would betray the doctor-patient relationship and destroy the sacred trust implied in that relationship. Consequently, the patients' medical conditions would be compromised. At the same time, the standing of the medical center (and, by extension, the medical community) would be irretrievably damaged in the eyes of patients.

This latter argument carried great weight with the administrators at the university. Baltimore made an argument along similar lines, namely, that to give data to those who were not "duly constituted" would "severely disrupt ongoing scientific activities."[62] The ability of, not just Baltimore and his team, but the entire scientific community to produce beneficial knowledge would be damaged if that community was not left to police itself.[63] In both arguments, the practitioners in each community should be left in charge of their respective products. But the university's position was the stronger since the product in question was the welfare of patients, not scientific knowledge.

Requests for access to the patients began almost immediately following the *Washington Post* story. As early as October 11, 1971, Saenger refused a local reporter such access on the ground that it was unfair to subject them to interviews in light of the "great attention to the fatal aspects of far advanced cancer."[64] (He failed to mention that it was the director of the medical center, Gall, who was the one who brought "great attention" to the patients by repeatedly referring to them as fatally ill during the press conference.) In early December, during an interview with Saenger and Silberstein, two members of Kennedy's staff, Ellis Mottur and Dr. Philip Caper, requested interviews with the patients to find out "the way consent methods are used and how they work." Since the five adults and three children who were still alive should have received this type of consent, Kennedy's staff believed that they could ascertain the quality of the consent process. Saenger immediately took the request as an affront and, according to his own notes, demanded to know "whether there was some question as to the integrity of Dr. Silberstein and himself." In his terms, the request was an offense since, as physicians, Saenger and Silberstein should be taken at their word when they said that they had fully informed the patients. We can appreciate from Saenger's response the importance of authority and trust within the medical community and why neither the Suskind nor the ACR committees had attempted to interview the patients. During further exchanges, Saenger and Silberstein also argued that Mottur and Caper were not qualified to interview patients and that they could not do so ethically.[65]

Following the meeting, Kennedy sought his own experts and, in a letter to Warren Bennis, the university's president, stated that, after discussion with a number of "authorities in medical ethics," it was, he felt, appropriate for the Senate to directly communicate with the patients. Kennedy further argued that the adults and the parents of the children "should have the opportunity to make up their own minds."[66] In a series of internal letters, Saenger outlined his many objections to Kennedy's request. Interviews would serve "no purpose in the care of their [the patients'] illnesses," and they could prove "very distressing." Kennedy's staff was not qualified to and could not fairly or ethically interview the patients. Saenger even suggested that a panel of expert physicians might interview the patients as well as a comparable group at another institution "so as to constitute a valid scientific inquiry." He was arguing that the consent process at the University of Cincinnati should be judged, not according to whether it fulfilled the regulatory requirements, but according to how it compared to national practices. He also raised the danger of adverse publicity if the administration agreed to Kennedy's request since "it would be comparatively easy for a sophisticated interrogator to convey an impression to the public that these patients had been grievously exploited." But his most salient argument was in allegiance to the liberal profession of medicine. The medical center could not accede to Kennedy's request for the names of the patients, Saenger claimed, "if we are to continue to merit the responsibility and trust which our patients manifest in us."[67] The medical community, and, in particular, the university medical center, could not continue to function if it did not retain the authority and trust of the community. This argument carried much weight since it rallied many in the medical center behind the administration. Even the contentious Suskind Committee took time from its internecine warfare to unanimously declare that it was "opposed to any outside investigating body interviewing a patient" since it "would violate the patients' rights."[68] In spite of internal support and the Ohio senator Robert Taft's work on behalf of the medical center, the university administration was under enormous political pressure to reveal the names of the patients.[69] Mottur continued to leak statements about upcoming hearings and the threat of cutting off funds to the university.[70] In spite of the political power of Kennedy's committee, the university administration turned down his request in mid-December, citing Ohio statute law and court decisions on a "patient's right to protection from public and resultant mental suffering."

The university's stance that it was protecting patients was attacked in a follow-up letter in mid-January in which Kennedy claimed that Cincinnati physicians had previously released the names of patients to the staff of public television and that tapes of the interviews showed the patients had not been properly informed.[71] In a long and convoluted letter, Saenger attempted to explain to Gall why he had failed to mention the previous interviews. He admitted that he had "erred," but he again turned to the argument of the prerogatives and responsibilities of physicians and the sharp boundary between medicine and research. Although he conceded that information "obtained by Federal funds should be made available both to the scientific and lay public," he also claimed that "the part of our medical therapy having to do with direct care . . . seems to me to be within the judgment of the attending physician." Kennedy had a right to have access to his notebooks and to the scientific data since the government financed the study (here Saenger parted company with Baltimore), but he did not have a right to intervene in the clinical care of patients. Saenger was arguing that his products, the patients' therapies, were not subject to outside authority but, rather, within the patient-physician relationship. And the sacred duties of the physician trumped Kennedy's demand for access and transparency.[72]

Gall circulated a draft response to Kennedy arguing that the distress following the public television interviews only strengthened the university's resolve to continue to refuse "uncontrolled probing by Subcommittee investigators," which would "contribute grievously to their [the patients'] emotional and physical detriment."[73] Saenger, however, cautioned the medical center to rethink a flat refusal because of the adverse publicity.[74] The administration took his advice, and Gall sent a more conciliatory letter to Kennedy informing him that they were polling the patients to find out how they felt about Senate interviews and requesting a meeting between Kennedy and senior members of the university administration.[75] The administration also sought further support from the medical community and consulted two physicians from the American College of Physicians about whether the Senate interviews should be permitted.

The response of the patients to the poll was unanimous; none of them wanted anything to do with Senate staff. A parent of one of the patients stated: "I wish no interview with the Senate, concerning my son's health. I believe this is very personal to my family and all the Doctors involved."[76] The patients shared with the physicians a vision of a world of medicine as protected from outside intervention. It was a position that, as we saw in chapter 5, their children would not share in their testimony during the 1990s. The two physicians from the American College of Physicians also supported the medical center and defended the liberal profession of medicine. They were against subjecting the patients to interviews that would lead them to "lose faith in the competence and motivation of the physicians" and "would be definitely deleterious" to their "physical and emotional status."[77]

The governor of Ohio, John Gilligan, held a meeting on February 24 that brought together Kennedy and senior members of the university administration.[78] It marked the end of Kennedy's demand for access to the patients. By that time, he had received (redacted) copies of the patients' responses and the recommendation of the American College of Physicians, in addition to the Suskind report, the ACR report, and a recent General

Accounting Office investigation that had cleared the university of improper use of federal funds. He left the meeting determined to seek national legislation on informed consent, while Bennis proclaimed that the basic issue was not so much due process, as in law, but "due care of patients."[79]

His attempt to gain access to the patients brought Kennedy into direct confrontation with the prerogatives of physicians, which the university administrators vigorously supported since their own viability was at stake. They understood that, if they had agreed to release the names of the patients, they would have undermined their position within the institution. The adverse publicity following patient interviews could have been damaging in its own right, but the medical center and its leaders could have survived the political fallout. What the administrators might not have politically survived was the likelihood that their own medical faculty would have viewed their behavior as a capitulation to outsiders and a betrayal of the faculty and the trust of their patients.

Throughout the battle, the welfare of the patients, while in principle primary, was often secondary in practice. Recall that Saenger had willingly given the names of patients to public television and that he had offered to enlist a patient for Rapoport. But the secondary status of the patients was, perhaps, most dramatically conveyed in a series of interviews that Silberstein held with all the adult patients in early January. The excerpt that follows gives much of those interviews' flavor:

> S: We have decided under no circumstance to give your name to any investigator from Kennedy's health sub-committee . . . that you have a right under the Constitution if someone should get your name . . . you have the perfect right to say I don't want to talk. Or you have a perfect right to say whatever is on your mind. But nobody can threaten you into talking. Do you understand that?

> M: Yes, I understand that.

> S: We have in discussing this decided . . . that it would be medically unwise . . . even if we wanted to give out your name . . . which of course we don't, we couldn't. You are protected by Ohio State law against that. . . . We also wanted to be sure that there was no doubt in your mind as to the uses of the specimens that we obtained from you, that the specimens of blood and urine would be used for research to help people who had received radiation either accidentally from one of these power reactors . . . or in the case of warfare either civilian bombing or of soldiers in the battlefield and you were told this also before you were treated. Am I correct?

> M: You are correct.[80]

Silberstein sent a handwritten note to Gall on January 12 indicating that he had interviewed all the adult patients. "All of them," he stated, "do not wish to talk to any Senate investigators. In my interviews with them I did not counsel them against talks." He concluded: "All but one patient remembered that some of our research results could be used to help soldiers in the battlefield as well as civilian casualties."[81] When he received Silberstein's note, Gall was in the midst of his exchanges with Kennedy over access to the patients. In a letter to Kennedy shortly afterward, Gall could confidently propose to poll the patients since he already knew the results.

A Phoenix Rises from the Ashes

How did Saenger's research program, which was subjected to continuing criticisms for methodological weaknesses and ethical difficulties, receive approval from the FCR, ACR, and Suskind committees and continue to receive the support of his many collaborators? And what issues led to its ultimate demise? Before turning to this issue, however, I need to sketch the events that led to that demise.

In early February, prior to the public release of his report, Suskind wrote to Bennis and recommended that Saenger's study should not continue in its present form. However, if the investigators wished to continue, Suskind argued, the prior study could be expanded into a randomized trial comparing chemotherapy and TBI for a select group of patients (those with advanced cancer of the colon). Since the DOD would not be interested in funding such a large-scale therapeutic project, Suskind suggested that the investigators would need to seek other financial resources. He, however, had no objection to the DOD continuing to support the metabolic and physiological components of the study.[82] Gall too was in favor of continuing the DOD arrangements.[83] Suskind's and Gall's positions make it evident that they, along with the Suskind Committee, continued to believe that a boundary had and could continue to be maintained between the therapeutic and the military components of the program. The study could meet its military aims without influencing how the patients were treated.

At about the same time as Suskind's letter, and at the height of the controversies, Saenger informally negotiated a new contract with the DOD. Although Gall and Grulee were aware of his discussions, they warned him not to reach any formal arrangement.[84] Nevertheless, on April 21, Saenger agreed to a contract with the DOD. He later defended his action on the ground of his duty as the principal investigator: "Thus in the absence of indication or direction of University policy prior to April 21, 1972, I regarded it as my responsibility to make every possible effort to secure immediate support for our staff."[85] When Bennis learned about the agreement, he issued a directive immediately ending all arrangements with DOD.

Two contingent events framed this story. First, Saenger's earlier decision to send an investigative reporter his DOD reports clearly abetted (if it did not initiate) public disclosure. Second, his April agreement with the DOD precipitated Bennis's action to end all DOD funding. Bennis may have been leaning in that direction and used Saenger's actions as an excuse, or he may have taken the advice of Suskind and Gall regarding continuing DOD support. Whichever way Bennis was leaning, Saenger's indiscretion forced him to issue his directive because of the possibility of adverse publicity and even another confrontation with Kennedy. Remember that Kennedy had, for all intents and purposes, dropped his investigation following the meeting with Governor Gilligan. There is, however, an indication in Saenger's notes that Bennis assured Kennedy that the university would not go forward with a renewal with the DOD without notifying him first.[86] If this is so, Bennis's hand was clearly forced by Saenger, whose action, in this light, appears even more foolhardy.

Although the DOD funding had ended, the Saenger-Silberstein program was not moribund. Saenger negotiated interim support for the TBI program from the university administration following the DOD debacle,[87] while, in early April, and prior to Bennis's directive, Silberstein submitted a proposal to the FCR for a randomized trial comparing TBI and chemotherapy.[88] As in the previous submissions, the proposal went through a complex and troubled history. It took four months until the FCR approved the proposal (in its fourth version) and only, according to Hess, after "long exhaustive (and occasionally exhausting) review."[89] The investigators submitted the study to the NCI in September, and Saenger's program finally ended with the NCI's recommendation not to fund it.

Why did the TBI program survive so long? How was it possible that it continually renewed itself after each challenge? If we follow the program from 1966, when Friedman first submitted it to the FCR, the proposals appeared in three major incarnations, all criticized by the committee members on various grounds. Overall, the FCR had to review (by my reckoning) thirteen revisions, and any careful reading of them would lead to the conclusion that in no case were the committee's objections substantively addressed by the investigators. In addition, recall that Friedman's proposal had been rejected by the NCI in the mid-1960s on ethical grounds. In spite of these difficulties, the project was approved by the FCR three times (once provisionally), and it received a generally supportive reading by the Suskind

Committee. The ACR report, though tainted by McConnell's overly supportive rendering, must still be taken as evidence of the general support of the committee's three experts. McConnell could not have written such a splendid appraisal if the committee had denounced the program as poor science or unethical medicine. He could not have turned a swayback into something resembling a thoroughbred.

Why did the project continue to survive? We saw in chapter 6 that it had already become deeply embedded in the workings of institutional life, having been taken up by various laboratory and clinical investigators throughout the medical center. It is not surprising, then, that the FCR and the Suskind Committee had members who were either collaborators on the TBI project or colleagues of collaborators with, thus, an indirect interest in the program. When the *University of Cincinnati News Record* questioned the integrity of the Suskind Committee report, it charged that six of its eleven members either collaborated on Saenger's research, received funding from the TBI program, or were involved in earlier FCR review committees.[90] On the most mundane level, it would have been in the personal interests of some of the university staff to support the TBI program. If they turned it down, some of them might have damaged parts of their own research programs.

However, self-interest alone is not enough. There were other, more basic resistances to rejection. First, most of his reviewers did not believe that Saenger's proposals were pathological entities residing, as it were, in some universe of unethical practices. Rather, they viewed his studies as being of the same kind as their own and inhabiting a common community of medical research. If they rejected his proposals, they would have (at least in part) been questioning their own scientific and ethical probity. Second, since the program was not outside common practices, the reviewers felt no strong compulsion to reject it; rather, they could discipline and reform and even expand it—a position that was consistent with their belief in the gradual accumulation of knowledge through research. Experimentation was, in this view, fraught with errors, miscalculations, and wrong turns. Yet errors could be exploited to improve and even expand research efforts. This scientific ethos is powerfully expressed in Suskind's February letter to Bennis. In two places, Suskind makes the same point, that Saenger's study should be disciplined but not rejected: "In [the committee's] deliberations there was agreement that the *Whole Body Radiation Project should be continued* but that the study in its present form *should not*." Suskind was self-consciously aware that he was framing his recommendation in terms of *continuing* the project rather than (the starker alternative) rejecting it and beginning anew. As he put it, continuance "would allow the Project to effectively use its past experience to develop conclusive information." And what was that experience? "All of the achievements of the current study, such as autologous bone marrow transplants and intensive psychological support."[91]

Suskind's position (as well as that of the FCR) to discipline—rather than reject—the TBI project might, alternatively, be seen as an expression of a kind of self-renewing dynamic of scientific research. Failures justify more, rather than less, intervention. Research cannibalizes its own errors. McConnell's ACR defense of Saenger is another striking example of the turning of failure into a rationale for a further extension of research. Recall that McConnell had addressed the question of whether some eight patients had died of radiation sickness. He argued that more studies were needed since it was not possible "to determine positively that those patients who died within 60 days . . . would not have succumbed to their disease." What better place to turn, he urged, than Saenger's own program. McConnell quoted his committee's support for Saenger's argument that, since the small numbers of patients "preclude any claim of therapeutic superiority," it "seems reasonable to continue therapy for these gravely ill individuals."[92] He took the question about whether TBI caused death, transformed it into a query about TBI's "therapeutic superiority," and used the lack of knowledge of the latter to justify expanding the program.

There are two related and interwoven arguments that have emerged in this chapter, one dealing with the TBI program, the other with the governance of the medical center. The first strand of the story followed Saenger's program as it—chameleonlike—changed character in the hands of different actors. It was not only that various investigators took it in different directions but also that peer reviewers read it in their own ways and forced it to change under scrutiny. The more the program became entangled among various groups, the more it was buoyed up, and the more difficult it became to disengage it from its surroundings. The self-interests of its supporters and the belief in the expansion of science all kept the program from running aground. At the same time, the entire network that constituted the program was not a static enterprise. The project began and ended in response to local contingencies. It started gradually, through Saenger's DOD contacts, and ended more rapidly, following a series of rash acts and Saenger's decision not to contest the NCI's rejection of his last proposal. The second strand of the story followed the medical center's battles with the press and, especially, the U.S. Congress. In these encounters, the center's allegiance to the liberal profession of medical/scientific practices and its desire to continue to regulate its own clinical and research practices sustained it in the face of intense public scrutiny. Its striking ability to face down the powerful U.S. Senate speaks strongly to how much was at stake for the university administration (and the medical community). The story also emphasizes that the argument about protecting the welfare of patients provided the medical center with a trump card in its battles with the government.

Although, by the fall of 1972, Saenger's TBI program had come to an end, the affair surrounding it remained unresolved. The institution's ambiguous stance left Saenger and his coworkers, as well as his many critics, unsatisfied. Saenger spent the next years writing articles in his own defense and trying to recover from the trauma, while critical reports appeared sporadically.[93] The affair had not reached closure and would be revisited two decades later when various attempts were made to construct Saenger's program as a paradigmatic and pathological case.

Notes

For simplicity, files from the ACHRE folder DOD 042994-A; file no. /16 are written thus: DOD file no.

1. "Pentagon Has Contract to Test Radiation Effects on Humans," *Washington Post*, October 8, 1971.

2. "Kennedy to Investigate Radiation Test Project," *Washington Post*, October 9, 1971.

3. *Congressional Record* 117, no. 154 (October 15, 1971): S16371–S16372.

4. Osborne, "On Liberalism," 349. Osborne discusses medicine in the context of liberalism and neoliberalism from the viewpoint of governmentality, a constellation of concepts that was developed in the later work of Michel Foucault. For a more general discussion, see Dean, *Governmentality*; and the references in chapter 8.

5. The Warren Commission of course kept many of its records from public disclosure. This, however, only heightened the uncertainties and fueled the even more public debates, reports, books, and so on.

6. More specifically, a joint meeting between Oak Ridge Laboratories and the Defense Atomic Support Agency (DASA).

7. *Joint ORAU-DASA Information Exchange Program*, March 29–30, 1971, DOD 7.

8. Edward B. Silberstein, "The Political and Ethical Investigation of Human Research: A Case Study," n.d., 6, DOD 042994-A;5/16.

9. Saenger to Rapoport, "To give appropriate perspective . . . ," April 19, 1971, DOD 10.

10. Rapoport, *Great American Bomb Machine*, 83 (tables), 86–87 (Operation Priscilla).

11. Ibid., 90, 93 (see also 92–93). The "expert," Dr. John Gofman, of the University of California, Berkeley, was a noted critic of the environmental dangers of radiation. In *Radiation and Human Health*, he claimed that the regulations on background radiation led to much higher levels of cancers than were indicated in reports commissioned by the American and European radiology communities, which dismissed

Gofman's work. Whatever the merits of his analyses of radiation effects, Gofman's expertise was not in radiotherapy clinical trials.

12. Ibid., 90. A similar statement was made during this period to the effect that "what is good for GM is good for the country." It took General Motors almost three decades to live down that indiscretion.

13. Ibid., 96.

14. Boyer, *Promises to Keep*, esp. chap. 6. The link between science and the military clearly had other and earlier critics than the campus demonstrators of the late 1960s. C. Wright Mills's (*Power Elite*) argument against the military-scientific complex and President Eisenhower's often-quoted warning of 1961 about the "danger that public policy ... could become the captive of the scientific technological elite" are but two examples. See Mendelsohn, "Politics of Pessimism," 156.

15. Saenger to Rapoport, "To give appropriate perspective ...," April 19, 1971, DOD 10. Saenger's concern with waning federal support is revealing since Nixon's other war (on cancer) meant large increases in funding for medical research. Certainly, Silberstein's closed system for bone marrow transplantation would have fit well into a war on cancer. But, for Saenger, it was a zero-sum game; more funding for cancer meant less for atomic warfare questions. His mind, if not his heart, appears to have remained on the nuclear battlefield.

16. ACHRE, *Final Report*, chap. 8.

17. *Washington Post*, October 8, 1971.

18. *Paris L'Express*, October 18, 1971.

19. University of Cincinnati Public Information Office, "List of out of town media ...," October 11, 1971, DOD 8. See also E. B. Silberstein and E. L. Saenger, "The Political and Ethical Investigation of Research: A Case Study," n.d., 10, DOD 5.

20. "Summary of Events, Whole Body Therapeutic Radiation Study," n.d., DOD 3. This document appears to be one prepared for university legal counsel ca. January 1972.

21. Merton, "Normative Structure of Science"; and Taylor, *Defining Science*.

22. Edward A. Gall, "Statement Delivered by Dr. Edward A. Gall," October 11, 1972, DOD 050294-A;3/16.

23. Official Release of the University, October 11, 1971, DOD 12.

24. Opening Statement of Edward Gall, Press Conference Held at Cincinnati General Hospital, October 11, 1971, DOD 3 (hereafter cited as Press Conference).

25. Saenger Note, "A Statement in Regard to Whole and Partial Body Radiation Therapy ...," October 11, 1971, 1, DOD 8.

26. The text of the press conference contains an important addition: "(The work did not start until 1960.)" The fact that this statement is enclosed in parentheses makes it likely that it was added later. Although it could have been meant to indicate a break in Saenger's response, such a parenthetical insertion appears nowhere else in the text.

27. Opening Statement of Edward Gall, Press Conference.

28. Saenger Note, "A Statement in Regard to Whole and Partial Body Radiation Therapy ...," October 11, 1971, 4, DOD 8.

29. Press Conference, 4 (Gall quote), 10 (Grulee statement).

30. Ibid., 8, 12, 13.

31. Ibid., 7.

32. Opening Statement of Edward Gall, Press Conference.

33. Saenger Note, "A Statement in Regard to Whole and Partial Body Radiation Therapy...," October 11, 1971, 4, DOD 8.

34. Gall to Endejann, "Re attached inquiry...," October 25, 1971, DOD 8.

35. McConnell to Gravel, "This letter represents...," January 3, 1972, DOD 1 (hereafter cited as ACR Report).

36. Junior Faculty Association, "A Report to the Campus Community," ca. January 1972, DOD 1 (hereafter cited as JFA Report).

37. University of Cincinnati Ad Hoc Committee, "To the Dean of the Medical College by the Ad Hoc Review Committee of the University of Cincinnati," January 1972, DOD 1 (hereafter cited as Suskind Report).

38. Porter, *Trust in Numbers*, 149.

39. As early as October 21, less than two weeks after the *Washington Post* story, Gravel had written to the surgeon general seeking answers to his many concerns, which included questions like: Was there any "trickery" of the patients? The surgeon general refused to respond and had one of his surrogates dismiss Gravel's questions with a brief statement to the effect that Saenger conducted valid studies. The American Cancer Society and the NCI followed suit. Gravel to Steinfeld, "Subject: Trickery," October 21, 1971, DOD 1; and O'Connor to Whom It May Concern, "Re: Development of Saenger Case," May 15, 1972, DOD 060194-A.

40. Linton to Gravel, "As indicated in his letter...," November 24, 1971, DOD 12.

41. ACR Report, 2. The committee consisted of Drs. Henry Kaplan, Frank Hendrickson, and Samuel Taylor. For their impressive credentials, see ibid.

42. Ibid., 1, 3.

43. Ibid., 3, 9 (use of DOD funds), 13.

44. Gravel to McConnell, "It is clear from your letter...," February 4, 1972, DOD 3.

45. McConnell to Gravel, "After receipt of your letter...," March 7, 1972, DOD 10.

46. During congressional hearings in the mid-1990s, one of Saenger's critics argued that McConnell was a "long time fishing partner" of Saenger's and had "conducted the review." "Statement of David Eligman, House Judiciary Committee, Subcommittee on Administrative Law and Government Relations," April 11, 1994, 9, DOD 042994-A;13/16. Although it was the committee of three experts that interviewed the investigators and communicated its findings to McConnell, a careful reading of the ACR report, or, more precisely, the letter to Gravel that was written and signed by McConnell alone, leaves little doubt that it was laundered.

47. "Statement of Martha Stephens, House Judiciary Committee, Subcommittee on Administrative Law and Government Relations," April 11, 1994, 1 (quote), 2, DOD 042994-A;13/16.

48. JFA Report, 4.

49. "Statement of Martha Stephens, House Judiciary Committee, Subcommittee on Administrative Law and Government Relations," April 11, 1994, 2, DOD 042994-A;13/16.

50. JFA Report, 2.

51. Grulee to Gall, "On January 28, I was asked...," January 31, 1972, DOD 1.

52. Grulee to Suskind et al., "I am most grateful...," November 12, 1971, DOD 1.

53. The report was actually written in a question-and-answer format as an aid for the university president.

54. Suskind to Members of Review Committee, "In the discussion with...," December 2, 1971, DOD 3.

55. "Continue Radiation Project, UC Advised," *Cincinnati Post*, February 2, 1972.

56. Suskind Report, 67 (the favorable result for colon cancer), 35 (the psychology section apparently having been lifted), app. A (the history). The psychology section contained descriptions of the studies using, e.g., *we* and *us* rather than *they* or *the investigators*. That section was the responsibility of the chairman of psychiatry (a committee member), and the material appears to have been inserted verbatim into the report. Hess's account is misleading in a number of places compared with the archival record.

57. Ibid., 46, 61, 66 (recommended expansion), 46.

58. Saenger to Grulee, "Enclosed is my last letter...," February 9, 1972, DOD 3.

59. See Chubin and Hackett, *Peerless Science*, 138-53; and Kevles, *Baltimore Case*, 67-95.

60. Guston, *Between Politics and Science*, 92-98.

61. Chubin and Hackett, *Peerless Science*, 144.

62. Ibid., 145.

63. Steven Shapin makes a similar point: "The demands for accountability appeared [in the twentieth-century United States] radically incompatible with the autonomy that, scientists said, was the condition for the health of science, its capacity to yield objective knowledge, and thus, to produce knowledge on which technological innovation could be placed" ("Science and the Public," 1004).

64. Saenger to Moores, "In regard to your request...," October 11, 1971, DOD 8.

65. Saenger Memorandum, "An Interview with Dr. Silberstein and Mr. Mottur and Dr. Caper...," December 6, 1971, 6 (quote), 7, DOD 3.

66. Kennedy quoted in Saenger to Barrett, "I have reviewed the letter...," December 17, 1971, DOD 3.

67. Ibid.; and Saenger to Gall, "You requested an opinion...," December 11, 1971, DOD 11.

68. Suskind Report, 56.

69. Iaft to Kennedy, "Thank you for your letter...," December 23, 1971, DOD 11.

70. "Doctors Say UC Patients Not Guinea Pigs," *Cincinnati Post*, January 4, 1972. Mottur is quoted in the article to the effect that Kennedy may subpoena university and medical center officials. See also "Kennedy Aide Says UC Refusal Endangers Funds for Research," *Cincinnati Enquirer*, January 4, 1972.

71. Kennedy to Bennis, "This is in reply to your acknowledgement...," January 11, 1972, DOD 050294-A;3/6.

72. Saenger to Gall, "This letter will comment on...," January 13, 1972, DOD 10.

73. Gall Draft Letter to Kennedy, "Your letter of January 1, 1972...," n.d., DOD 11.

74. Saenger to Gall and Barrett, "In regard to the answer to Mr. Kennedy...," January 14, 1972, DOD 10.

75. Gall to Kennedy, "This is in reply to your letter of January 11...," January 19, 1972, DOD 050294, 3/6. The polling took place through letters to the patients that read: "They want me [one of the physicians] to give them your name and tell them where you live, but I have refused to give them this information without your O.K." Silberstein to [patient], "A member of the United States Senate...," January 18, 1972, DOD 10; and Horowitz to Dear Mr. and Mrs., "A member of the United States Senate...," January 24, 1972, DOD 10.

76. Patient to Silberstein, "As I have stated before...," February 1, 1972, DOD 10.

77. Ross to Gall, "As you requested...," February 10, 1972, DOD 8; and Kennedy to Gall, "Enclosed is a brief statement...," February 14, 1972, DOD 8.

78. Bennis to Kennedy, "This is in response...," February 16, 1972, DOD 8.

79. University of Cincinnati News Record, "Gilligan Joins in Radiation Controversy...," March 7, 1972.

80. MM Interview with Silberstein, January 4, 1972, DOD 10.

81. Silberstein to Gall, "I have now spoken...," January 12, 1972, DOD 10.

82. Silberstein to Bennis, "In a recent conversation...," February 7, 1972, DOD 3.

83. Gall to Bennis, "The report of...," January 26, 1972, DOD 8.

84. Saenger to Grulee, "Enclosed is my last letter...," February 9, 1972, DOD 3; and Grulee to Gall, "I have checked with Gene Saenger...," February 9, 1972, DOD 8.

85. Saenger to Gall, "This letter is sent to you...," July 5, 1972, DOD 3.

86. Saenger Unsigned Note, "DNA advised by Kennedy staff and GAO that univ. would advise Kennedy's office of any further proposal if the univ. was going to accept it," April 24, 1972, DOD 3.

87. Following Bennis's directive, Saenger demanded that the institution pick up the project expenses for the period of a year unless interim funding could be found. After a series of charges and countercharges, Saenger was able to procure interim support from the university administration. Saenger to Gall, "Attached is the budget...," April 24, 1972, DOD 10; and "Recheck of UC Project," Cincinnati Post, April 25, 1972.

88. Silberstein to Hess, "Enclosed is a proposal...," April 4, 1972, DOD 3.

89. Hess to Grulee, "The Faculty Committee on Research...," August 28, 1972, DOD 3.

90. "Suskind Comm.—Conflict of Interest?" University of Cincinnati News Record, February 11, 1972.

91. Suskind to Bennis, "In a recent conversation...," February 7, 1972, DOD 3.

92. ACR Report, 8.

93. Bierwaltes to Saenger, "Your letter of May 8, 1975 makes me realize for the first time what a nightmare...," May 19, 1975, DOD 10.

PART III

ATTEMPTS AT CLOSURE

8 Ethical Judgment

Resistance to definition sets the limit to sovereignty, to power, to the transparency of the world, to its control, to order. • Zygmunt Baumann, *Modernity and Ambivalence*

....

On January 15, 1994, President Bill Clinton created the Advisory Committee on Human Radiation Experiments (ACHRE) to investigate the history of government-sponsored radiation experiments during the cold war. The committee came into being very rapidly as the result of continuing public disclosures about unethical experiments with radiation on human subjects. The furor had begun a few months earlier following a series of reports in the *Albuquerque Tribune* about Americans who had been injected with plutonium during the cold war.[1] By early December, Hazel O'Leary, the head of the Department of Energy (DOE), held a press conference at which she referred with shock to stories about radiation experiments on citizens. She announced that she had hired an ethicist to look into the allegations, that the cold war was over, and that "we're coming clean."[2] O'Leary was referring to what had become a deluge of sensational news exposés about cold war radiation experiments and the abuse of the public. In Massachusetts, for example, there were charges that retarded students had been fed oatmeal cookies laced with small quantities of radioactive materials. There were stories from the Northwest about Atomic Energy Commission-funded research involving the irradiation of the testicles of prison inmates. The 1986 U.S. House of Representatives investigation of the so-called American nuclear guinea pigs was resurrected, as was the Saenger affair. The Cincinnati papers dredged out the stories from the 1960s at about the same time as Saenger received the prestigious Gold Medal of the Radiological Society of North America in 1993.

Clinton responded rapidly, creating ACHRE to investigate human radiation experiments between 1944 and 1974, the year when the government issued standardized rules for protecting subjects.[3] The committee was chaired by the ethicist Ruth Faden and consisted of two additional ethicists, five physicians, two lawyers, two scientists, one historian, and one "citizen representative," a bank vice president. By the time it was in full swing, the committee had a staff of nearly seventy. It was charged with determining the ethical standards of the period of time under investigation and whether the experiments were consistent with those standards. It was also to consider whether the experiments had clear medical or scientific purposes and included appropriate medical follow-up. It might also "recommend further policies, as needed, to ensure compliance with recommended ethical and scientific standards for human radiation experiments."[4]

Eighteen months later, the nine-hundred-page report and various supplements submitted by ACHRE covered an impressive range of material. The first section provided an overview of the ethical practices in research with human subjects. The second part contained nine chapters of case studies covering cold war radiation experiments with plutonium, nontherapeutic research on children, the total-body irradiation (TBI) experiments (not only in Cincinnati, but also in Oak Ridge, Houston, New York, and Bethesda), radiation experiments on prisoners, atomic bomb test exposures of soldiers, intentional releases of environmental radiation, and other experiments. The third part of the report presented the results of the committee's research into contemporary practices. The final part contained the committee's twenty-three findings and eighteen recommendations and covered well over forty double-column pages.

In spite of this major effort, the committee could not reach a clear and consistent judgment in the Saenger case, a difficulty that had plagued previous reviewers. As we have seen, beginning in the mid-1960s, some of Saenger's peers at the University of Cincinnati had voiced their ethical concerns about his project and genuine fears for the safety of the patients, but, in spite of those concerns, his studies were never rejected as unethical or dangerous. The situation was not much different for the numerous internal and external review committees in the early 1970s. They too could not reach clear and consistent judgments about the ever-changing TBI program, and the one critical voice—that of the Junior Faculty Association (JFA)—was simply ignored. Consequently, none of the committees could pin down the changing research enterprise long enough to judge it and, perhaps, stop it. In the end, a contingent event forced the university president to end contact arrangements with the Department of Defense (DOD), but even he did not take a clear stance on the TBI studies.

At first glance, we might have expected ACHRE to reach a definitive and withering judgment. The committee had a formidable ethical machinery to bring to bear on Saenger's program. The committee's chair, Ruth Faden, had coauthored with Thomas Beauchamp a central text in bioethical thought, *A History and Theory of Informed Consent*, and her coauthor had coauthored one of the field's canonical texts, the *Principles of Biomedical Ethics*. The latter text, the foundation of so-called principlism, presented a set of ethical principles—patient autonomy, beneficence, nonmaleficence, and justice—that have become common currency in our day. These principles were first developed in 1979 as part of the U.S. government–commissioned *Belmont Report*.[5] That report, which was a product of a national commission created in 1974 by Congress to investigate research with human subjects, was highly influential and provided the legal groundwork for the ensuing government regulations.[6] ACHRE also had various bioethicists on the committee itself and a number of others among its numerous support staff. In addition to Faden, the committee included Ruth Macklin, a staunch defender of the role of principles and their universal applicability, as well as Jay Katz, a psychiatrist and longtime critic of medical researchers who has written extensively on human experimentation and informed consent. The support staff included Allen Buchanan, who, among other things, authored a position paper on retrospective moral judgments (see below), and Jonathan Moreno, who has written extensively on informed consent and postwar government-sponsored experimentation. For all its internal differences, the bioethics community had problematized the closed paternalistic world of medicine and demanded that it be opened itself to scrutiny from above and below (by ethical regulators and patients) and held accountable. According to the bioethical credo, the practices of the medical community would be governed by the strict demands of rational judgment and its universal ethical principles.

If ACHRE had marginalized Saenger and his ilk, if it unambiguously denounced the TBI program as the epitome of unacceptable practice, then it would clearly demonstrate the difference between normal and unethical behavior. If it branded Saenger a pathological case, one among only a small number of such examples, then it could rightly argue that ethical rules were basically working and that there was a need for only incremental changes. New and improved rules would provide improved bioethical control of research, and these rules would come from principles that had been codified following the *Belmont Report*. But it was not only its ethical machinery and its desire to support the continuing centrality of bioethical assessments that would lead one to expect ACHRE to reach a definitive judgment. Most important, it had extensive evidence in hand, certainly more than any of the previous groups that had looked at Saenger's program, and it was charged with acting as the voice of ethical reason. ACHRE thus occupied a commanding position from which to assess the TBI program and, if warranted, marginalize Saenger and his coworkers as paradigmatic of unethical research practices.

The reasons that ACHRE could not reach a definitive judgment in the Saenger case, ironically, arose in part from its reputed strengths and commanding position. To begin with, its access to an enormous number of resources proved, perhaps, as much a burden as a blessing. It had a large staff and budget, and, as part of a cabinet-level working group, it also had access to a huge archive that the DOE, the DOD, the Department of Health and Human Services, and other departments were mandated to locate and deliver. But the scope of the project that it faced was enormous, so enormous that the scale of the documentation surrounding the revelations of Saenger's experiments described in the last chapter seems small in comparison. The sheer quantity of documentation that ACHRE confronted was staggering, the individual pieces numbering well into the hundreds of thousands.[7] The committee was not entirely prepared to handle the endless cartons that arrived daily at its headquarters. Faden acknowledged early in the proceedings that she and her colleagues were overwhelmed. In addition, the scope of the committee's work included, among other things, a study of contemporary consent practices at each federal agency; 125 research projects were identified for closer scrutiny, including interviews with nineteen hundred patients. ACHRE also initiated an oral history project, held sixteen public meetings, each of two to three days' duration, and four field hearings, including one in Cincinnati, and listened to the testimony of more than two hundred public witnesses.[8]

One of the major problems that ACHRE faced was how to take control of the staggering amount of material in its possession. The volume of documents placed serious demands on its resources, so much so that it did not fully annotate its collection and chose to organize its acquisitions by provenance alone. Although it did produce a limited aid for searching through its collection, it did not have, as the DOE did, the time or the resources to provide online access to its valuable archive.[9] In addition, the sheer volume of resources stymied their full interrogation. For example, because the documents came from so many sources, the archive was swamped with multiple versions of one and the same memorandum or report—sometimes ten or more—ranging from drafts, to notes and comments (some handwritten), to photocopies, some complete, some incomplete. We saw in an earlier chapter how patients' families bitterly complained about the state of the their relatives' records. Of course, historical excavations require resolving such problems. Yet the vast scale of the materials placed a severe limitation on how many of the myriad questions the texts raised could be answered by even a group of researchers. Indeed, handling such archives is clearly a problem for much of postwar historical studies. Even so, as ACHRE argued in its report, not all the relevant material could be acquired.[10] Still, the sheer bulk of what was, in fact, collected restricts the uses to which the material can be put.

For example, since a limited amount of time can be devoted to individual documents, the archives must be interrogated differently than a small collection. In my own work, I quickly realized that one way into the archives was to focus on a single actor or a single set of theoretical issues and then go through the material to see what it yielded. Through trial and error, some of my ideas inevitably fell away to be replaced by others, but the size of the archive meant that I could attempt only a limited number of such trials. Little is said in the final report about how ACHRE handled these difficulties, although one early passage remarked that, within a few months, after self-education and a review of a few cases, "the outlines of a world that had been almost lost began to reemerge."[11] The situation, however, was certainly more complex. ACHRE investigators had to bring their own categories to the archives in order to impose some sense on them. One has only to read through its interviews to appreciate the emphasis the committee placed on, for example, informed consent, a notion that is central to contemporary bioethics. Moreover, the ACHRE staff appears not to have had the time to study all the available material, at least for the Saenger case, and this had important consequences for the conclusions reached, as we will see below.

There was, however, a more fundamental reason than the mountain of materials why no definitive judgment was reached regarding the Saenger case (and why it has remained a contentious issue for so long)—namely, that the case demonstrated the difficulty involved in characterizing and, consequently, judging research studies in terms of prescriptive rules. As we will see below, even the question whether Saenger was doing standard therapy, research, or some hybrid of the two was difficult to assess. Before considering this issue in more detail, however, we first need to briefly look at the changes in governance that occurred between the termination of Saenger's studies in the early 1970s and the compilation of the ACHRE report of the mid-1990s.

Changes in the Governance of Medical Research

The governance of medical research went through significant changes between the period of Saenger's TBI program in the mid-1960s and the point when ACHRE published its report on human radiation experiments in 1996. An appreciation of those changes—which included the rise and dominance of bioethics as a cultural and political voice in judging medical practices—is important to an understanding of ACHRE's handling of the Saenger case. Here, I want to concentrate on providing a characterization of contemporary bioethics and then indicate the influence that changes in the political culture had on earlier methods and aspirations.[12] To begin with, we can appreciate some flavor of the complex and contentious character of bioethics by looking at it through three distinct points of view: those taken in David Rothman's *Strangers at the Bedside*, an influential and highly quoted account by a medical historian, Albert Jonsen's *Birth of Bioethics*, an insider history written by a prominent bioethics practitioner, and Renee Fox's *Sociology of Medicine*, a scathing sociological critique. Rothman is strongly influenced by the 1960s liberal perspective, particularly its disdain for medical paternalism, Jonsen, as a casuist, lies somewhat outside the mainstream of fundamentalist bioethical thought, and Fox, an early participant in the bioethics movement, has in recent years become an intensely partisan critic of many of that movement's positions and practices. Yet their various arguments provide a sense of the continuing controversy surrounding bioethical issues and some of the strengths and weaknesses of bioethical positions, particularly principlism and its place in the broader context of U.S. history.

In *Strangers at the Bedside*, Rothman lays out some of the major themes of postwar medical practices. According to his account, prior to World War II, and through the mid-1960s, medicine was governed by individual decision making. Physicians reached ethical and clinical judgments on a case-by-case basis and were, for the most part, not subject to outside scrutiny. By the mid-1980s, medicine had been decentered, and normative medical ethics had been imposed by outsiders (Rothman's *strangers*)—that is, government officials, health care administrators, ethics experts (bioethicists), and the like. Individual decision making had given way to institutional review boards (IRBs) and discussions of risks, costs, and benefits, while more formalized practices and extensive written documentation replaced word-of-mouth medical orders. Medicine was no longer private, or a matter for the medical profession alone, but a public concern expected to display an openness and transparency typical of other institutions. Finally, Rothman located the fulcrum of change in the highly charged revelations of unethical experimentation that became the subject of public hearings during the 1960s. One of the signifying events for him was, as I mentioned in chapter 2, a 1966 exposé in the *New England Journal of Medicine* by one of the field's prominent researchers, Henry Beecher, who recounted a number of unethical medical experiments.[13] Rothman also emphasized the significance of the high proportion of indigent and black populations in such experiments. He claimed that this aspect of medical experimentation provided an important rallying point for reformers during the turbulent 1960s.[14]

In contrast, most bioethicists—and, in this sense, Jonsen's history offers a typical account—assume a deterministic role for technological change. It was medical technology that had led to new and unprecedented claims on ethics, and history and tradition had provided neither the public nor physicians with sufficient understanding to solve these ethical dilemmas. For Jonsen, then, the fulcrum of change should be located not so much in the unethical treatment of, say, syphilitic blacks from 1932 to 1972 at the Tuskegee Institute in Alabama (for all the madness of that sorry affair) as in the appearance of new technology. The birth of bioethics should be sought in cases like the introduction of renal dialysis, which forced the Seattle Artificial Kidney Center in 1962 to come to terms with an unprecedented ethical dilemma, namely, how to fairly deliver a scarce medical resource. Similarly, the Karen Quinlan case was yet another ethical problem caused by new technology. In 1975, Quinlan was brought comatose to a New Jersey emergency room, where she was placed on life support equipment, but her neurological condition deteriorated. Her parents demanded that the attending physicians disconnect her from life support and allow her to die, but they refused. The case was settled in 1976 when, on the basis of Quinlan's right to privacy, the New Jersey Supreme Court ruled that the physicians could remove her from life support without legal liability.[15] From Jonsen's bioethical viewpoint, the development of machines for artificial respiration had forced physicians and patients to confront unprecedented ethical problems. Moreover, the elevation of these incidents to "cases" signified for Jonsen their central importance. For most bioethicists, then, such new technologies have changed our ideas about the nature of medicine and led to the following questions: Who should have control over the new technologies? Who should make moral decisions? How can individuals be protected? What is the fair distribution of new technologies?[16] These questions exemplified new categories of ethical dilemmas in which issues like justice and autonomy demanded answers that only a new philosophy of bioethics could deliver.[17]

For all the sensitivity and nuances of their renditions, Jonsen and Rothman produced primarily triumphalist accounts of moral progress. For Rothman, the old order of paternalistic physicians was swept away (or, better, displaced) by coteries of outside experts in medical governance. Although the title of his book makes clear that these changes were not without cost, Rothman has little sympathy for paternalistic medicine. For Jonsen, the technological imperative required a new breed of ethical experts (in his case, casuists) who could apply rational philosophical arguments to answer ethical questions that were beyond the ability of physicians and patients alike.

In contrast, Renee Fox, who recognized and applauded many of the changes that occurred in the governance of medicine, nevertheless took the new bioethics movement to task for its weaknesses and its blindness to its own shortcomings. Beyond a broad agreement with the other critics over key events, Fox was scathing in her account of the bioethics movement and the ethos that supported it. She argued that bioethics had mistakenly accorded primary status to individual rights, autonomy, and self-determination and that it had negatively characterized paternalism, no matter how well-meaning, as a blatant injustice because it interfered with individual freedom. She also argued that, in bioethical accounts, the relationship between physician and patient was dominated by the notion of contracts, relationships whose operative ethical modalities were "self-conscious, rational, functionally specific agreements between independent individuals" and whose archetype was informed consent. This emphasis on an overtly liberal agenda based on contracts between patients and physicians overshadowed, according to Fox, issues of social responsibility, obligations, and duties. Furthermore, the rational thinking so highly prized by bioethicists led to a largely deductive, abstract, and formal enterprise in which the human context was simplified, if not entirely removed. All concerns were seen as transnational and universal, and the local context of problems was either dismissed or given short shrift. According to Fox, the bioethics movement was deeply conservative: not only was it dominated by its belief in individualism, but it also relied on a discourse of middle-class professional academics and scholars who had little appreciation of the social embeddedness of medical/ethical problems. Thus, as Fox points out, the biomedical issues that arose, for example, from the social conditions of an underclass, like the neonatal problems faced by women in poverty, were seen as outside the bioethical consensus and off the agenda.[18]

What were the views of the bioethics movement itself? How could it govern medical practices? To begin with, bioethics was anything but a single unified movement; it was, and still is, a congeries of medical/ethical programs under a single banner.[19] Among the plethora of contending ethical theories are virtue ethics, casuistry, and communitarian, pluralistic, feminist, and narrative ethics, to name but a few. In addition, the term *bioethics* has been applied to a range of biomedical issues that extend beyond medical ethics per se, most notably to issues of environmentalism and animal rights. Yet one variety of bioethics has dominated the agenda and is known under various sobriquets: *principle-based ethics*, the *four-principle theory of ethics, fundamentalism*, or, for short, *principlism*. Not only are a great majority of bioethics texts based on a principlist approach, but also, and more important, it forms the basis of many government and hospital-based ethical rules and practices.

In the canonical text on principlism, *Principles of Biomedical Ethics*, Beauchamp and Childress have expounded the four-principle approach.[20] According to Childress, principlism "remains the most influential framework in bioethics."[21] The importance of principlism for the Saenger case is that it provided the bioethical foundation supporting ACHRE's deliberations and *Final Report*. The first edition of *Principles* was published in 1979 (the same year as the *Belmont Report*), and the ideas that informed it were similar, though not identical, to those in *Belmont Report*. Also, as mentioned previously, in 1986 Ruth Faden coauthored with Beauchamp the canonical text on informed consent, *A History and Theory of Informed Consent*. The ethical theory that informed the *History* was nearly the same as that informing the *Principles*, except the number of principles had been reduced from four to three. ACHRE took over this earlier work, increased the number of ethical principles to six, and made them the basis of the committee's deliberations.

According to Beauchamp and Childress, ethical decision making should be based on their four principles. The first addresses liberal rights (the principle of autonomy), the second doing no harm (the principle of nonmaleficence). The last two refer to benefits, risks, and costs, and address either their utility (the principle of beneficence)[22] or their fair distribution (the principle of justice). These principles are meant to have no intrinsic hierarchy or ordering, the bioethicist giving each its proper weight for the case at hand. They are intended to provide a framework in which rational thought will lead to closure, even in intractable cases like Saenger's. The principles are derived from "socially approved norms of social conduct" and, in combination, "put common morality as a whole into a coherent package."[23]

Principlists have, however, been criticized, not only for their claim to know the "socially approved norms," but also for their lack of appreciation of the pluralism of moral views within and across national boundaries.[24] Principlists usually respond by acknowledging pluralism as a social reality, but, nonetheless, they insist on the underlying universality of their principles. For example, although ACHRE recognized that there were plural views on moral issues within American society, it claimed that its principles were widely enough shared that the plural positions would converge to the views of the committee. There was, in effect, a "public endorsement" of the moral claims made "by official bodies charged to speak for society as a whole," and that included ACHRE.[25]

Yet one of the problems ACHRE had in its deliberations was balancing an inevitable tension within the bioethics movement itself. On the one hand, bioethicists had, from the beginning, taken a radical position arguing for patients' rights and an end to medical paternalism. On the other hand, by the 1990s bioethics had gained enormous power and prestige and had assumed a central role in the management of medical research, sometimes at the government level. As we will see below, ACHRE's *Final Report* was not immune from this strain between radicalism and governance. To understand the source of this dilemma, we must move beyond strictly bioethics-based accounts and appreciate that the movement arose during the political changes that occurred over the course of the Saenger affair, namely, the neoliberal transformation of American politics and society that led from the earlier liberal welfare state (a consensus forged by President Franklin Roosevelt in the 1930s) to the mid-1980s conservative revolution of the Reagan presidency.[26] By viewing bioethics, not simply as an agenda of 1960s liberals, but as a program that shared in the broader neoliberal movement, we can begin to appreciate the problem of the double alliance that confronted the bioethics movement (typical of neoliberal programs),[27] namely, the alliance to both political authority and the individual.

In the 1980s, the old political system that had been forged by Roosevelt and that had reached its apotheosis under Lyndon Johnson was attacked by neoliberals as bankrupt from a number of directions. For example, neoliberal critics claimed that, instead of bringing improved standards of living to the indigent, the welfare agenda had engendered feelings of apathy and nonparticipation in liberal democracy and the market. Moreover, neoliberals attacked not only the disappointing results of liberal welfare programs but also the means that the state used to try to achieve its aims. The welfare state was increasingly controlled, the argument went, by government bureaucrats, experts who dictated how society was to be administered. Their power had spread well beyond welfare and social programs and was endangering the freedom of people to participate in an open market economy. Large-scale welfare programs were constantly attacked for the contaminating influence of the centralized bureaucracy in Washington.

The professions—and medicine is an important example—were located by neoliberals within the same liberal welfare state problematic. Medicine was a closed and hierarchical profession run by experts who claimed to know what was best for patients and society. Neoliberals argued, instead, for the "empowering" of individuals and communities. Consumerism became a central dogma of neoliberalism and contributed to the dismantling of the liberal welfare state in the course of the 1980s and 1990s. It formed the basis of a new technology of politics demanding the consumer's right to choose, whether it was the brand or pedigree of produce in a supermarket or a new experimental therapy for cancer.[28]

Bioethics and neoliberalism rose to importance and power contemporaneously. Bioethics itself can be viewed as a complex movement that shared some of the new ideas of neoliberalism while maintaining practices of the older liberal welfare agenda. On the one hand, the bioethical stand against paternalism and its support of the rights of patients fit readily within the neoliberal political technology of consumerism and empowerment. On the other hand, bioethics retained and participated in the paradigm of a liberal welfare state run by experts. In many respects, bioethics was co-opted by the state (and, to a lesser extent, by the medical profession) and made a voice of power and, through its universal principles, found itself opposed to, or at least not in active support of, a more pluralistic ethics of communities. Bioethicists became, not the outsiders of Rothman's story, but consummate insiders, facilitators of medical research, not only within individual hospitals, but also in any number of large-scale government programs.[29] As an embodiment of contemporary bioethics, ACHRE, as we will see below, also faced a similar problem, namely, how to fulfill its role as an arm of political authority and still maintain its support for the rights of individuals and the plurality of community positions.

ACHRE's Judgment of Saenger's Program

In was in the context of these developments that ACHRE attempted to judge postwar human experimentation. Its Final Report, which was (nearly) unanimous on the surface, was also fragmented. Although the report argued for a grand vision for the future role of bioethics, it could not in its case studies produce definitive judgments of the cold war researchers. In part, ACHRE's deliberations were affected by its effort to apply universal principles retrospectively, which led to a split in the committee. Indeed, almost from the outset, the committee was divided over the issue of retrospective judgment. The ethicists and some of the support staff argued that ethical assessments using universal principles could and should be made about earlier practices and, in particular, about individual investigators, while the dissenters, for example, professionals from medicine and law, adopted various alternate positions. For some, ethical judgments across time were neither possible nor appropriate, especially if the committee were to identify and blame particular individuals. Others felt that, even if universal judgments could be made in theory, the standard of evidence necessary to hold individuals accountable could not be met. The ethicists took varying positions, but they settled on the argument that, if individuals were not held accountable to universal principles, the committee's report could not hope to deter future researchers from participating in unethical practices.[30]

The ethicists sought to answer the dissidents' concerns by developing a system of ethics in which wrongdoing and culpability were clearly separated. In this schema, Saenger, for example, could be judged to have harmed patients but, owing either to what ACHRE termed *culturally induced moral ignorance* or to insufficient evidence, not be held culpable. The framework for retrospective judgment consisted of three kinds of ethical standards: basic principles, government policies, and professional rules of ethics. The ethicists characteristically argued that basic principles were universal and applicable to judgments across time and space but that government policies and professional rules of ethics were not because they had historical contexts. These layers would then provide a basis for balancing wrongdoing and culpability. The (universal) basic principles were nothing more than those in the earlier bioethical texts, the *Belmont Report*, *Principles of Biomedical Ethics*, and *A History and Theory of Informed Consent* in an alternate and expanded guise. There were six principles—three "ought-nots" (treat people as mere means, deceive others, and inflict harm) and three "oughts" (promote welfare, treat people fairly, and respect self-determination).

But the solution of retrospective judgment that the committee adopted to try to close the fissure between the ethicists and the dissidents only opened the door to further dissent. Separating blame and culpability provided an opening that the dissenters could use to blunt the ethicists' attempt to impose universal judgments. The dissenters argued that it was not possible for the committee to single out individuals as blameworthy since it had not investigated the records in sufficient detail to levy such judgments. Moreover, there were concerns among the dissenters about judging experimenters who were no longer alive and, thus, unable to defend themselves.[31] For the ethicists, these were disingenuous arguments that masked the dissenters' inability to effectively argue about the principles of retrospective judgment. Allen Buchanan, who wrote the position paper on retrospective moral judgment, complained that some of the ACHRE members acted in bad faith. Although the dissenters had signed on to the retrospective judgment paradigm, Buchanan believed that they undermined it with objections about standards of evidence once they had lost the ethics debate.[32]

Buchanan did not, however, appreciate the dilemma that faced the dissenters and the committee as a whole once the machinery for retrospective judgment had been constructed. The physicians on the committee certainly appreciated the complexities and moral ambiguities that are such an intrinsic part of professional practices, and, in this respect, the universal principles that the ethicists were pushing were simply too blunt an instrument to bring to the table. Had the six principles been invoked to condemn Saenger, they could just as readily have been turned on, for example, the physicians themselves. And the dissenters' resistance to rendering universal judgments would not have been attenuated by the bioethicists' further claim that they could characterize an investigator as blameworthy but not culpable. Indeed, it would be hard to imagine that physicians would embrace the position that, during the cold war, so many of their own had suffered from culturally induced moral blindness. The fight here was between two rationalities: a universal one that could be brought to bear on all cases and one underwritten by a fear that, if anyone was guilty, then all might be.

At the same time, Faden was faced with another issue, namely, blending ACHRE's role in governance with its support of the rights of the individual. On the one hand, she believed that the voice of the victims should not be silenced. "There's nothing more terrifying for survivors of a horrible event," she had stated, "than to hear other people trivialize it, or even worse, raise skepticism about whether an event ever occurred." In many respects, the report was able to realize this agenda. It documented the many abuses that the radiation subjects had suffered, and the committee provided sympathetic forums for the families of many of the victims. Yet, as mentioned above, in the *Final Report* we do not hear the individual voices of the human subjects. On the other hand, she also argued that her aim was to do nothing less than "rewrite the history of ethics and research on human subjects in this country."[33]

Here, the committee faced a significant challenge, that of Jay Katz, a persistent critic of medical experimentation. He was bothered by ACHRE's insider argument that research practices had substantially improved since the cold war and that more of the same governance would make the control of research practices even better: "The present regulatory process is flawed. It invites in subtle, but real ways, repetitions of the dignitary insults which unconsenting citizen-patients suffered during the cold war." For Katz, a more root-and-branch approach was required to protect citizen-patients.

ACHRE's program was part of the problem, not the solution.[34] ACHRE's argument, however, was that medical research had greatly improved since the cold war, which supported its claim that the ethical regulations that had evolved since the *Belmont Report* were working and that only changes around the margins were required to further improve medical research practices. Still, some of the changes had far-ranging implications.

Many of ACHRE's recommendations addressed well-circumscribed issues, for example, the conditions under which subjects should receive notification and/or apologies from the federal government (another battle within ACHRE).[35] They also addressed specific ways in which IRBs could be improved. The far-ranging nature of ACHRE's goals is, however, best exemplified by its ninth group of recommendations, which argued for nothing less than the undertaking of a program "on a national scale to ensure the centrality of ethics in the conduct of scientists whose research involves human subjects." What ACHRE was seeking was a "mandate" for "the teaching of research ethics" in order to overhaul the "culture of human subjects research." Bioethics would govern such research by accustoming the medical community to its universal ethical precepts. Moreover, these changes were to be brought about through a national advisory panel or commission that would develop and, more important, "perhaps implement" a series of far-reaching recommendations. ACHRE sought to extend the requirement of offering programs in the "responsible" conduct of research to all federal grant recipients. It also recommended requiring all medical students, house staff, and fellows to take courses in research ethics. It likewise sought a national program in which the leaders of biomedical research would spearhead efforts to elevate the status and role of ethics. The last of its recommendations, however, was the most far-ranging. It sought to require the ethical oversight of all American medical accrediting bodies (like the Joint Commission on the Accreditation of American Hospital Organizations), a move that could potentially bring all medicine (not simply research on human subjects) under the oversight (and, thus, control) of ACHRE's national ethics commission.[36]

In the context of these various battles and political goals, ACHRE produced an analysis of Saenger's research program. It addressed some important issues, among them: whether Saenger favored research over therapy; whether TBI contributed to the death of some of the patients; and whether his research practices met the standards of his peers. In addressing these questions, the committee was overall unable to produce a definitive ethical judgment. On the first issue, it presented evidence to support the conclusion that there was evidence that demonstrated "the subordination of the ends of medicine to the ends of research." Specifically, it argued that, since Saenger sought to understand the physical effects of TBI on healthy soldiers, he withheld giving emetics to his patients prior to treatment (as well as avoiding full ethical disclosure) since these actions might adversely affect the experiment. At the same time, the committee felt compelled to add the following caveats: "*To the extent* that this [lack of disclosure, emetics] deviated from standard care and caused unnecessary suffering and discomfort, it was morally unconscionable; *to the extent* that the standard of care in this area is uncertain, it is morally questionable" (emphasis added).[37]

The committee also pointed out that the cohort Saenger treated with TBI consisted almost entirely of patients who had so-called radioresistant tumors, rather than radiosensitive ones. This was important, it argued, because patients "with radiosensitive tumors (for which TBI was considered most promising) were less useful subjects for obtaining . . . information on the acute effects of radiation on healthy soldiers and citizens." If Saenger had treated radiosensitive patients, he would have found it more difficult to assess "whether signs of such nausea, vomiting, or other acute effects were due to the rapid destruction of cancer cells by radiation or due to the radiation acting on normal tissue, such as normal blood cells." Similarly, patients with radiosensitive tumors were, for comparable reasons, also less useful for finding (or constructing) biological dosimeters. Yet, in spite of this, the committee also appreciated that a therapeutic case could, in fact, be made to treat radioresistant tumors for palliation of symptoms (using higher doses, possibly accompanied by bone marrow transplantation), even though it would have been "nonstandard therapeutic practice . . . at that time." Finally, the committee was most troubled by evidence that Saenger would not have begun or continued using TBI were it not for DOD funding. In interviews with committee staff members, Saenger stated that, if he had found a biological dosimeter early on, he would have gone on to do something else.[38] But Saenger also argued that the funds were necessary since they provided people, laboratory equipment, and a protocol "to look at whole body radiation in comparison with other forms of palliation." Although none of these issues could be unequivocally resolved, the committee at least concluded (in a weaker form) that "the impact of the research protocol on the care of the patient subjects cannot be construed as beneficial to the patients" or, as I already mentioned (in a stronger form), that there was evidence that medicine was subordinated to research.[39]

While the committee had concluded that Saenger's program favored research over therapy, on the question whether TBI contributed to the early deaths of some of the patients it was simply unable (or unwilling) to provide a definitive answer—even though it had stated that "it sought to determine what effect, if any, the DOD requirements had on the actual treatment of patients."[40] Not only was this issue important for the committee's ethical assessment of Saenger, but it also had considerable meaning for the families of some of the patients, among them Maude Jacobs's family. In its *Final Report*, ACHRE argued that it was simply unable to look at the record in sufficient detail: "Although the Advisory Committee has received some partial patient hospital records, it has not analyzed the records of every patient, which would be required to determine if any deaths could be attributed to TBI alone[41] or if such conclusions could be reached at all from the data currently available. The Committee did not have the time or resources to review individual files of every patient from this and the numerous other experiments that it has investigated."[42]

Wrongdoing, at least according to this section of the report, could not be determined in the face of a lack of resources. Instead, the committee revisited the contradictory positions of the Suskind and American College of Radiology (ACR) committees as well as some of Saenger's writings about whether TBI contributed to the deaths of some of the TBI patients. The Suskind Committee claimed that it had identified nineteen patients whose early deaths could possibly have been caused by radiation alone even though marrow failure was found in only eight. Yet it too concluded: "There is absolutely no evidence that whole body radiation shortened the period of survival of the treated patients." The ACR was also equivocal. While it associated eight deaths with subnormal marrow function "relatable to radiation syndrome," it also argued that "it is not possible to determine positively that those patients who died within 60 days of treatment would have succumbed to their disease within that period." Saenger himself took equivocal and contradictory positions. In a 1973 publication, he suggested that, if marrow death *could be* attributed to TBI alone and not to cancer or previous treatments, then "one can identify 8 cases in which there is a possibility of the therapy contributing to mortality." In a 1994 report, he had evidently backed away from this earlier position since there he argued: "It is important to realize that in any given patient it is not possible to determine objectively whether death occurred too soon or was prolonged as a consequence of treatment."[43] Finally, the one unambiguous assessment, a statistical analysis by the JFA—which held that Saenger's treatments led to an increased number of short-term deaths—was not included in the *Final Report*.[44] All the positions on patient deaths that ACHRE discussed were presented without comment. The committee did not weigh the evidence and left unanswered whether TBI itself killed some of the patients, whether, on balance, the TBI program decreased survival chances, or even whether such questions could be answered.

ACHRE's difficulty in achieving a clear assessment of Saenger's program is, however, most apparent in its attempts to compare his practices to those of his contemporaries in order to determine whether they met the standards of the day. It did so by first partitioning medicine into exclusive categories and then placing Saenger's program into one of them. To make clear the difficulty faced by the committee, I trace its arguments here in some detail. It began by parsing the world of medicine into three categories: "In the practice of medicine there has always been a fine boundary between practices or treatments that are accepted as standard, those that are 'innovative,' and others that are experimental or the subject of research." The quotation marks were no doubt meant to point out that the category *innovative* was not well defined and that it should include those medical interventions that were neither strictly research nor strictly therapy. The category paid heed to the fact that so many new ideas are first tried in the clinic outside formal research protocols. But the addition of this third category only increased the difficulty. Indeed, the committee stated: "By this time [the mid-1960s], total body irradiation was not standard treatment for such cases, nor could it be called innovative treatment; some at the time considered its use in patients with radioresistant cancer to be controversial." It seems that Saenger's study fell into, not any of the previous categories, but a new, fourth category, *controversial.* The committee, however, immediately reversed itself and implied that TBI may, indeed, have been innovative therapy: "The history of medicine is replete with instances in which failure [of an innovative therapy] is followed by success."[45] It seemed to be saying: How can we judge Saenger? Perhaps his controversial (innovative?) therapy might have paid off after all.[46]

At this point in the report, immediately following the quote just given, ACHRE produced the following double-negative judgment of Saenger's program: "The continued use of TBI in patients with radioresistant cancers *would not* have been *unethical* if the physicians had established clear benchmarks for determining how much additional use was warranted, and if patients have been informed of the speculative nature of the treatment and the gravity of the risks involved" (emphasis added). The committee's difficulty in this concluding section is palpable. We would, it almost seems to conclude, not even argue that the TBI studies should not have been done, although we would have expected to see some additional constraints. Yet, as the committee confessed, it could not even conclude whether those constraints had been met: "It is not clear that either of these things [the stopping rules and informed consent] occurred." Once it moved on from the specific to the general, the fog lifted. In the very next paragraph, the ambiguous and uncertain tone immediately changed to certainty: "What *is clear* is that neither the university's IRB nor the funding agency reviewed the appropriateness of continuing to treat patients with radioresistant cancers using TBI without bone marrow protection" (emphasis added). By this point, the committee was more than able to mete out blame and culpability. Indeed, after some further remarks along the same lines, it concluded: "The responsibility for failure rests at all levels."[47] Ambivalent individual judgments had been followed by definitive and sweeping generalized pronouncements. The situation was not unique, being repeated throughout the case studies. Tom Beauchamp, in a critique of the ACHRE report, noted: "Nowhere in the Advisory Committee's *Final Report* is a named agent (other than the federal government and the medical profession) ever found culpable."[48]

In one sense, the difficulty that ACHRE had in judging Saenger may have stemmed from incomplete medical records as well as from issues internal to the committee: a lack of time (or interest) to fully interrogate the records (specifically, the patient charts), a broad agenda that required many human resources, and even the effects of dissent. In another sense, however, the lack of ethical closure touches deeper issues. This becomes especially clear when we realize that the committee did try to categorize Saenger's practices, to locate them within the standards of the day and try to ethically judge Saenger in relation to his contemporaries. Indeed, the committee was careful and often prescient in its analysis of Saenger's research program. Yet it was unable to produce an unambiguous judgment. This was not simply the result of incomplete archival records or a lack of time and resources. It was because it had difficulty defining the research in question and pinning down a dynamic enterprise.

Indeed, the dynamic character of medical research has been a persistent theme throughout this study. We have seen that programs with multivalent goals flourish in large networks. We have also seen how research programs are constantly redefining their goals, for example, in order to align themselves with the next round of funding proposals. When such research projects confront a monitoring program like bioethics, that program becomes but one more node in the network of alliances that constitutes research enterprises. Investigators are adept at adjusting and refining their protocols and consent statements to meet or respond to an ethical evaluation by an IRB. An important component of such an assessment is the categorization of the therapy and its comparison to community standards; in some cases, these assessments can be quite difficult. And categorization and comparison become even more perplexing when the assessment is retrospective. We have observed that Saenger's enterprise had changed so much over a period of a decade that ACHRE could never clearly conclude whether he was doing research, or therapy, or some hybrid. I am not suggesting, however, that prescriptive ethical rules are worthless or that we should not try to develop and refine them. I am, rather, pointing out that, because of their dynamic character, research enterprises are sometimes difficult to categorize, with the result that prescriptive rules are limited in what they can accomplish.

In places in its report, ACHRE recognized the difficulty involved in categorizing medical practices. For example, it appreciated that the clinical objectives of the Cincinnati treatments remain "difficult to categorize even now." It also appreciated that there is a diffuse boundary (or, as the committee put it, a fine line) between research and therapy and "innovative" practices. Nevertheless, it never really made the connection that problems categorizing research practices and difficulty applying prescriptive rules might be related. Its subsequent recommendations supported a strong program in which medical research could be well defined and governed through more of the same type of rules. In particular, it concluded: "The history [of the military experiments] provides compelling evidence of the importance of the rules that regulate research today—prior review of risks and potential benefits, requirements of disclosure and consent, and procedures for ensuring the selection of subjects."[49]

ACHRE's account of Saenger's research was, however, anything but the last word on an affair that had already lasted more than three decades and drawn numerous individuals into its web. Saenger still faced a civil suit brought by family members of his patients as well as the judgment of some of his harshest critics, to which I briefly turn in the epilogue.

Notes

1. I rely heavily here on ACHRE, *Final Report*; Faden, "Chair's Perspective"; and Welsome, *Plutonium Files*.
2. Quoted in Welsome, *Plutonium Files*, 424.
3. The Department of Health, Education and Welfare issued the regulations.
4. ACHRE, *Final Report*, xxiv.
5. USDHEW, *Belmont Report*.
6. Specifically, the National Commission for the Protection of Human Subjects of Biomedical and Behavioral Research.
7. ACHRE, *Final Report*, xxvi; Welsome, *Plutonium Files*, 449; and Faden, "Chair's Perspective," 217. Welsome (*Plutonium Files*) claimed that there was something like

6 million pages of material, and Faden ("Chairs Perspective," 217) referred to the documentation that the federal agencies produced as "mountainous."

8. ACHRE, *Final Report*, xxvii–xxviii.

9. ACHRE, *Final Report, Supplemental Volume 2*. It also seems to me that the neat provenance structure for finding records was not consistently maintained, inventories of the records seeming to depend more on the individual in charge of an acquisition than on an overall policy.

10. ACHRE, *Final Report*, xxvi.

11. Ibid., xxv.

12. For narrative accounts of the rise of bioethics, readers should consult Fox's *Sociology of Medicine*, Rothman's *Strangers at the Bedside*, and Jonsen's *Birth of Bioethics* (all three of which are discussed below) as well as McKenny's *To Relieve the Human Condition* and esp. Stevens's *Bioethics in America* and Evans's *Playing God*.

13. Beecher, "Ethics and Clinical Research."

14. Rothman, *Strangers at the Bedside*, 70–84, and "Ethics and Human Experimentation," 1195–98.

15. Jonsen, *Birth of Bioethics*, 254–57.

16. Callahan, "Bioethics," 249.

17. Jonsen, *Birth of Bioethics*, xvii–xviii.

18. Fox, *Sociology of Medicine*, 229–30 (quote, 229).

19. For an overview, see Grodin, *Meta Medical Ethics*; Singer, *Companion to Bioethics*; and King et al., *Beyond Regulations*.

20. Beauchamp and Childress, *Principles of Biomedical Ethics* (1979), 37.

21. Childress, "Principle-Based Approach," 61.

22. The *Belmont Report* did not use the term *beneficence* to mean that a physician acts solely for the benefit of the patient. Societal benefits and costs were included along with patient benefits and risk in a utilitarian arithmetic that remains hidden under the benign term *beneficence*.

23. Beauchamp and Childress, *Principles of Biomedical Ethics* (1979), 37.

24. For a critique of principlism's transnational ethics, see Baker, "Theory of International Bioethics."

25. ACHRE, *Final Report*, 117.

26. For neoliberalism and the more general subject of governmentality, see Dean, *Governmentality*; Rose, *Powers of Freedom*; and the essays, esp. those by Gordon and Foucault, in Burchell et al., eds., *Foucault Effect*. For the American scene, see Lash, *True and Only Heaven*; Bellah, *Habits of the Heart*; and Cruikshank, *Will to Empower*. A discussion of the differences between neoliberalism and neoconservatism can be found in Dean, *Governmentality*, 159–64.

27. Rose and Miller, "Political Power," 188.

28. On clinical trials as a mechanism for coping, see, e.g., Löwy, *Between Bench and Bedside*.

29. Stevens discusses these dilemmas in some detail. She quotes Daniel Callahan, one of the founders of the field, who became concerned about bioethics' role as a

facilitator and caustically remarked: "Tell us what you want to do and we'll tell you how to do it ethically" (*Bioethics in America*, 66).

30. Buchanan, "Retrospective Moral Judgment," 246–48; and Welsome, *Plutonium Files*, 460.

31. This position was taken by the lawyer Nancy King (Welsome, *Plutonium Files*, 460).

32. Buchanan, "Retrospective Moral Judgment," 245–48.

33. Quoted in Welsome, *Plutonium Files*, 448, 447.

34. ACHRE, *Final Report*, 543 (Katz quote), 543–48. See also Welsome, *Plutonium Files*, 460.

35. Macklin, "Disagreement, Consensus."

36. ACHRE, *Final Report*, 522–23.

37. Ibid., 253.

38. Ibid., 229, 242, 253. More troubling for me were Saenger's desire to develop homologous transplants, which (unlike autologous transplants) could have been in the interest of only the military and would clearly have carried much higher toxicity for the patients, and the administration of psychomotor studies (rarely mentioned by any of the critics) to debilitated and dying patients.

39. Ibid., 242, 253.

40. Ibid., 240.

41. In Saenger's case, ACHRE did not need to review all eighty-eight cases; eight might have been sufficient.

42. ACHRE, *Final Report*, 245. According to Gary Stern, an ACHRE staff member, it was not a question of resources; the committee was determined not to review any of the records of injury (Stevens, *Bioethics in America*, xv).

43. ACHRE, *Final Report*, 245.

44. Indeed, Stephens's arguments (in the 1971 JFA report) are still the most compelling, and, although they would need to be brought up to date (Stephens lacked some of the records), they remain unanswered. The committee may have excluded the report since it had concerns whether it represented the official position of the JFA. See Gary Stern and Jonathan Engel, Interview with Professor Martha Stephens, October 20, 1994, 39–43, ACHRE 021095-A.

45. ACHRE, *Final Report*, 251.

46. ACHRE, of course, had in other places assumed that Saenger was doing research; e.g., the committee had, as we saw, concluded that medicine had been subordinated to research. These differing perspectives further highlight the difficulty it had in pinning down Saenger's program.

47. ACHRE, *Final Report*, 251.

48. Beauchamp, "Looking Back," 261.

49. ACHRE, *Final Report*, 241, 251, 250.

EPILOGUE

Jarndyce and Jarndyce drones on.... It has been observed that no two Chancery lawyers can talk about it for five minutes without coming to a total disagreement as to all the premises. Innumerable children have been born into the cause; innumerable people have married into it; innumerable old people have died out of it. • Charles Dickens, *Bleak House*

....

For many of those who became ensnared in Saenger's research enterprise and its aftermath, the search for effective closure was critical. For the investigative reporter Eileen Welsome, who began it all with her 1993 exposé on postwar radiation studies in the *Albuquerque Tribune*, there was some solace in a general apology Clinton made when he accepted the report from the Advisory Committee on Human Radiation Experiments (ACHRE) in October of 1995: "With the president's apology, the day of reckoning had finally come for scientists, doctors, and the bureaucrats who schemed for decades to keep knowledge of the plutonium injections and other radiation experiments from becoming public."[1] For Martha Stephens, the rulings of the federal court judge Sarah Beckwith in a 1994 civil suit brought by the families of Saenger's patients provided some clarity in judging Saenger and the team at the University of Cincinnati. Remember that, as a young assistant professor of English literature at the university, Stephens had become embroiled in the Saenger controversy in the early 1970s. She had written the 1972 Junior Faculty Association critique of the experiments, which was, as we have seen, the most trenchant analysis of the deaths that occurred among Saenger's patients. Following these events, Stephens lost contact with the Saenger affair, and her "life proceeded in other directions."[2] In the mid-1990s, the revelations about human radiation experiments during the cold war, however, brought her back into the fray. Her reentry into the Saenger case almost twenty-five years later is a striking indication of how deeply it affected everyone involved and how much they all sought to find some measure of closure.

Stephens testified before ACHRE and became deeply involved in the civil suit. She researched the stories of the patients, whose names were by then public, and in 2002 published *The Treatment: The Story of Those Who Died in the Cincinnati Radiation Tests*, in which she recounted many of their stories—dedicating the book to their memory. She had more in mind, however, than telling the stories of the patients. Her purpose was nothing less than a far-ranging indictment that fully documented the "widespread collusion by government at several levels, an institution existing in the public interest, and a broadband of the medical community." Her approach was one of disclosure and denunciation, and, in page after page of the book, she makes clear her contempt, not only for Saenger and his collaborators, but also for many of the university administration and the lawyers for the defense. There are few gray zones in her account. In one telling vignette, she is put off by a greeting from the lead defense attorney (Joseph Parker) since she reads it as suggesting that they behave civilly "without discomforting each other and having feelings about such things [the civil suit]." There is no room in Stephens's world for anything like this, and she only hopes that she is "not that sort of person."[3]

If there are any heroes in Stephens's account, they are the young civil rights attorney Lisa Meeks, who worked on behalf of the families, and Judge Beckwith, whose opinions in the case were a "gift for common justice." In a 1995 ruling, Beckwith dismissed the physicians' argument that they were immune from liability as public servants since, according to Beckwith, they acted as "scientists interested in nothing more than assembling cold data to be used by the Department of Defense." Once Beckwith had characterized the experiments as purely investigational, without any therapeutic merit, she could turn to the Nuremberg Code with some assurance since it had been used to judge nontherapeutic experiments in Germany. For Stephens, Beckwith's ruling was a clear denial "of the defendants' contention that they had not used people against their will." For Stephens, this ruling provided some measure of closure.[4]

The civil suit did not, however, produce a legal judgment in the Saenger case since the attorneys reached an out-of-court settlement in May 1999. Most of the families received something like $50,000, a minuscule financial settlement by contemporary American standards.[5] The case had ended rather badly. There was much dissension among the families about the monetary settlement, and, perhaps most significant of all, the civil suit had not settled the issues of blame and responsibility. Saenger declined to apologize, and the only recognition of the plight of the patients was a small commemorative plaque that was placed at an out-of-the-way location on the hospital grounds.[6]

Eugene Saenger died on September 30, 2007. The obituaries that followed his death hardly settled the controversies surrounding the case. The *New York Times* focused primarily on the radiation experiments and presented an essentially critical perspective. It quoted two of Saenger's most vociferous critics, Dr. David Eligman, a clinical associate professor of community medicine at Brown University, and Martha Stephens. Eligman characterized the radiation experiments in the grandest of terms: "What has happened here is one of the worst things this government has ever done to its citizens in secret." Stephens was more direct; she argued that, if the Cincinnati physicians had told the patients, "You may die of this radiation," then "there would have been no experiment."[7]

The *Cincinnati Enquirer* presented a more nuanced account of the different perspectives surrounding the radiation experiments. It also painted a much more personal portrait of Saenger, one that made room for his many accomplishments. It quoted Stephen Thomas (a medical physicist who began working with Saenger in 1975) that Saenger was "very sensitive, but very strong, and he stood up well, with the strength he always had, under the pressures that he faced." Although Thomas appreciated that the experiments were "products of a different time," he also pointed out that Saenger "felt he had acted entirely within the realm of medicinal science as it was known at that time. He definitely felt that he had not overstepped any bounds, any ethics in the research." Edward Silberstein saw him as "a man of incredible vision, seeing what was important before anyone else saw it." Even Martha Stephens is quoted as professing "very mixed feelings" about Saenger's skills and involvement in the radiation experiments. The obituary went on to list Saenger's various awards and accomplishments, among them the Eugene L. and Sue R. Saenger Professorship in Radiological Sciences, created at the University of Cincinnati when he retired in 1987; the George Charles de Hevesy Nuclear Pioneer Award for outstanding achievement in nuclear medicine from the Society of Nuclear Medicine, also in 1987; the Gold Medal at the Radiological Society of North America in 1993; the American Roentgen Ray Society's Gold Medal for lifetime achievement in 1998; his efforts to pass a local tax levy to pay for indigent care at the university hospital and the Cincinnati Children's Hospital; and his consultation with the military following the Chernobyl disaster. However, the controversy over Saenger's radiation experiments is, perhaps, best epitomized by the opening epigraph of the obituary: "In some circles, Dr. Eugene Saenger was reviled as the leader of controversial Cold War–era human radiation experiments. But colleagues revered the Indian Hill physician, who died September 30 at the age of 90, as a pioneering scientist on par with Marie Curie."[8]

After more than forty years, the Saenger case has continued to escape effective closure. Saenger's peers from the mid-1960s through the early 1970s had ongoing concerns about his research, but they always sought to try to discipline his program, rather than putting a stop to it. The program survived scrutiny, in part, because his research looked and felt too much like most everyone else's. The mid-1990s ACHRE investigations likewise did not lead to an unambiguous judgment, for similar reasons. Neither the civil suit of the 1990s nor Saenger's death in 2007 has resolved the underlying problems. This lack of closure illustrates the importance of the Saenger case for understanding medical research practices. It provides us with a lens through which to focus on postwar medical research, revealing that such enterprises not only contained components that we would judge as normal but also sometimes contained elements that we would find egregious. If the egregious emerges from the Saenger case, so does the role of the military and how deeply it was intertwined with radiation therapy during the cold war. And the Saenger case is impossible to understand without coming to terms with how much the medical community struggled during the postwar period with the problem of ethical governance. I would also argue that the Saenger case should not be understood entirely as a problem of science and medicine during the cold war because it touches on issues that more generally pervade much of medical research. It dramatically brings to light the problematic character of experimenting on human subjects, especially given the inherent tension between the needs of the patient and the goals of the investigation. It strikingly illustrates how the highly fluid character of research constantly changes in response to contingent events, including both scientific and ethical review. But, perhaps most important, and certainly most troubling, the Saenger case reveals how the prescriptive rules governing clinical conduct can provide room for a questionable program to thrive. As I mentioned in the introduction to this book, Giorgio Agamben in *Homo Sacer* claims: "The physician and scientist move in a no-man's-land into which at one point the sovereign alone could penetrate."[9] It is a no-man's land for Agamben because it is governed by a set of laws essentially beyond the state. It is a world of divided loyalties, especially for physician-researchers, who must (imperfectly) balance the needs of the patients with the goals of the research. But, primarily, it is a domain that allows for the flourishing of science; consequently, the space provided for the normal is open enough that the egregious can, at times, find room to operate as well. If that were not the situation, the Saenger case would have ended a long time ago.

Notes

1. Welsome, *Plutonium Files*, 470.

2. Stephens, *Treatment*, xi.

3. Ibid., 260, 279.

4. Ibid., 244, 46, 252.

5. Welsome, *Plutonium Files*, 476.

6. Stephens, *Treatment*, 289.

7. "Eugene Saenger, Controversial Doctor, Dies at 90," *New York Times*, October 11, 2007, available at http://www.nytimes.com/2007/10/11/us/11saenger.html?.

8. "Pioneering Radiologist Dies," *Cincinnati Enquirer*, October 3, 2007 (available through http://nl.newsbank.com/).

9. Agamben, *Homo Sacer*, 159.

BIBLIOGRAPHY

Most of the primary documents (besides published manuscripts, reports, and conference proceeding) are contained in Record Group 42 at National Archives II, 8601 Adelphi Rd., College Park MD 20740-6000. The Advisory Committee on Human Radiation Experiments (ACHRE) collected this material from various sources, including the Department of Defense (DOD), the Department of Energy (DOE), and corporate (CORP) and independent sources (IND), e.g., the University of Cincinnati. The material is referenced according to ACHRE's provenance system: A DOD entry, e.g., would appear as DOD 042994-A;7/16, which refers to the source of the document (DOD), the date it was received by ACHRE (042994, or April 29, 1994), the batch code (A), the file number (7), and the total number of files in the batch (16) constituting the acquisition. Sometimes in the endnotes the system is abbreviated; this is indicated at the heads of the relevant notes sections. In addition, ACHRE has interviewed a number of individuals, and they are referenced as ACHRE Interview Project. I have also used some references from the Internet, primarily http://search.dis.anl.gov/ (the DOE's online archive) and http://www.gwu.edu/~nsarchiv/ (which contains the transcripts of ACHRE meetings). The DOE archive is not currently available owing to lack of funding (see http://www.energy.gov/about/5308.htm).

Abeloff, Martin D., Allen S. Lichter, John E. Niederhuber, et al. "Breast." In *Clinical Oncology*, ed. Martin D. Abeloff, Allen S. Lichter, John E. Niederhuber, and James O. Armitage. New York: Churchill Livingstone, 1995.

Advisory Committee on Human Radiation Experiments (ACHRE). *Final Report of Advisory Committee on Human Radiation Experiments*. Oxford: Oxford University Press, 1996.

———. *Final Report of Advisory Committee on Human Radiation Experiments, Supplemental Volume 2: Sources and Documents: Appendix*. Washington, DC: U.S. Government Printing Office, 1996.

Agamben, Giorgio. *Homo Sacer: Sovereign Power and Bare Life*. Stanford, CA: Stanford University Press, 1998.

Armitage, James O., Anne Kessinger, and Elizabeth C. Reed. "Bone Marrow Transplantation." In *Clinical Oncology*, ed. Martin D. Abeloff, Allan S. Lichter, John E. Niederhuber, and James O. Armitage. New York: Churchill Livingstone, 1995.

Bailar, John C., and Elaine M. Smith. 1986. "Progress against Cancer?" *New England Journal of Medicine* 314 (1986): 1226–32.

Baker, Robert. "The History of Medical Ethics." In *Companion Encyclopedia of the History of Medicine*, ed. Roy Porter and W. F. Bynum. London: Routledge, 1993.

———. "A Theory of International Bioethics: Multiculturalism, Postmodernism, and the Bankruptcy of Fundamentalism." *Kennedy Institute of Ethics Journal* 8, no. 3 (1998): 201–31.

Barnes, D. W., M. J. Corp, J. F. Loutit, and F. E. Neil. "Treatment of Murine Leukemia with X-Rays and Homologous Bone Marrow." *British Medical Journal* 2 (1956): 626.

Baumann, Zigmunt. *Modernity and Ambivalence*. Cambridge: Polity, 1993.

Beauchamp, Tom L. "Looking Back and Judging Our Predecessors." *Kennedy Institute of Ethics Journal* 6, no. 3 (1996): 252–70.

Beauchamp, Tom L., and James F. Childress. *Principles of Biomedical Ethics*. 1st ed. New York: Oxford University Press, 1979.

———. *Principles of Biomedical Ethics*. 4th ed. New York: Oxford University Press, 1994.

Beecher, Henry, K. "Consent in Clinical Experimentation: Myth and Reality." *Journal of the American Medical Association* 195, no. 1 (1966): 34–35.

———. "Ethics and Clinical Research." *New England Journal of Medicine* 274 (1966): 1354–60.

———. "Experimentation in Man." *Journal of the American Medical Association* 169, no. 5 (1959) 461–78.

———. "Human Experimentation—a World Problem from the Standpoint of a Medical Investigator." *World Medical Journal* 7 (1960): 79–80.

———. "The Powerful Placebo." *Journal of the American Medical Association* 159 (1955): 1602–6.

———. *Research and the Individual: Human Studies*. Boston: Little Brown, 1970.

Begg, Colin B., Paul P. Carbone, Paul J. Elson, and M. Zelen. "Participation of Community Hospital Clinical Trials: An Analysis of Five Years Experience in the Eastern Cooperative Oncology Group." *New England Journal of Medicine* 306 (1982): 1076–80.

Bellah, Robert N. *Habits of the Heart: Individualism and Commitment in American Life*. Berkeley and Los Angeles: University of California Press, 1985.

Bloor, David. *Knowledge and Social Imagery*. 2nd ed. Chicago: University of Chicago Press, 1991.

Boyer, Paul. *By the Bomb's Early Light: American Thought and Culture at the Dawn of the Atomic Age*. New York: Pantheon, 1985.

———. "Physicians Confront the Apocalypse: The American Medical Profession and the Threat of Nuclear War." *Journal of the American Medical Association* 254, no. 5 (1985): 633–43.

————. *Promises to Keep: The United States since World War II.* Lexington, MA: D. C. Heath, 1995.

Brandt, Allan M. "Polio, Politics, Publicity, and Duplicity: The Salk Vaccine and the Protection of the Public." In *Major Problems in the History of American Medicine and Public Health*, ed. John Harley Warner and Janet A. Tighe. Boston: Houghton Mifflin, 2001.

Brandt, Allan M., and Lara Freidenfelds. "Research Ethics after World War II: The Insular Culture of Biomedicine." *Kennedy Institute of Ethics Journal* 6, no. 3 (1996): 239–43.

Buchanan, Allen. "The Controversy of Retrospective Moral Judgment." *Kennedy Institute of Ethics Journal* 6, no. 3 (1996): 245–50.

Bud, R. F. "Strategy in American Cancer Research after World War II: A Case Study." *Social Studies of Science* 8 (1978): 425–59.

Burchell, Graham, Colin Gordon, and Peter Miller, eds. *The Foucault Effect: Studies in Governmentality.* Chicago: University of Chicago Press, 1991.

Bush, Vannevar. *Science, the Endless Frontier: A Report to the President*, Washington, DC: Office of Scientific Research and Development, 1945.

Byerly, H. "Explaining and Exploiting Placebo Effects." *Perspectives in Biology and Medicine* 19 (1976): 423–35.

Callahan, Daniel. "Bioethics." In *Encyclopedia of Bioethics* (rev. ed.), ed. Warren T. Reich. New York: Simon & Schuster Macmillan, 1995.

Callon, Michel. "Some Elements of a Sociology of Translation: Domestication of the Scallops and Fishermen of St. Brieuc Bay." In *Power, Action, Belief*, ed. John Law. London: Routledge, 1986.

Childress, James F. "Principle-Based Approach." In *A Companion to Bioethics*, ed. Peter Singer. Oxford: Basil Blackwell, 1991.

Chubin, Daryl E., and Edward J. Hackett. *Peerless Science: Peer Review and U.S. Science Policy.* Albany: State University of New York Press, 1990.

Collins, H. M. *Changing Order: Replication and Induction in Scientific Practice.* Chicago: University of Chicago Press, 1985.

Congdon, C. C. "Experimental Treatment of Total-Body Irradiation Injury: A Brief Review." *Blood* 12 (1957): 746–53.

Cooney, James P. "The Physician's Problem in Atomic Warfare." *Journal of the American Medical Association* 14, no. 9 (1951): 634–36.

————. "Psychological Factors in Atomic Warfare." *Radiology* 53 (1949): 104–7.

Cox, James D. "Brief History of Comparative Clinical Trials in Radiation Oncology: Perspectives from the Silver Anniversary of the Radiation Therapy Oncology Group." *Radiology* 192 (1994): 25–32.

————. "Evolution and Accomplishments of the Radiation Therapy Oncology Group." *International Journal of Radiation Oncology, Biology, Physics* 33, no. 3 (1995): 747–54.

Craver, L. F. "Tolerance to Whole-Body Irradiation of Patients with Advanced Cancers." In *Industrial Medicine on the Plutonium Project: Survey and Collected Papers*, ed. Robert S. Stone. New York: McGraw-Hill, 1951.

Cronkite, Eugene P. "The Diagnosis, Treatment and Prognosis of Human Radiation Injury from Whole-Body Exposure." In "Physical Factors and Modification of Radiation Injury," ed. L. D. Hamilton. Special issue, *Annals of the New York Academy of Sciences* 114 (1964): 341–49.

Cronkite, Eugene P., and George Breecher. "Effects of Whole Body Irradiation." *Annual Review of Medicine* 3 (1952): 193–214.

Cruikshank, Barbara. *The Will to Empower: Democratic Citizens and Other Subjects.* Ithaca, NY: Cornell University Press, 1999.

Curran, William J. "Current Legal Issues in Clinical Investigation with Particular Attention to the Balance between the Rights of the Individual and the Needs of Society." In *Psychopharmacology: A Review of Progress 1957–1967* (Public Health Service Publication no. 1836), ed. Daniel H. Efron. Washington, DC: U.S. Government Printing Office, 1968. Reprinted in *Ethics in Medicine: Historical Perspectives and Contemporary Concerns*, ed. Stanley Joel Reiser, Arthur J. Dyck, and William J. Curran (Cambridge, MA: MIT Press, 1977).

———. "Governmental Regulation of the Use of Human Subjects in Medical Research: The Approach of Two Federal Agencies." In *Experimentation with Human Subjects*, ed. Paul A. Freund. London: George Allen & Unwin, 1970.

Day, Barbara. "The Medical Profession and Nuclear War: A Social History." *Journal of the American Medical Association* 254 (1985): 644–51.

Dean, Mitchell. *Governmentality: Power and Rule in Modern Society.* London: Sage, 1999.

Department of Energy (DOE). *Human Radiation Experiments: The Department of Energy Roadmap to the Story and the Records.* 1995. http://tis.eh.doe.gov/ohre/roadmap/.

Dessauer, F. "Eine neue Anordnung zur Röentgenbestrahlung." *Archiv für physikalische Medizin und medizinische Technik* 2 (1907): 218–23.

DeVita, Vincent T., Jr. "Breast Cancer Therapy: Exercising All Options." *New England Journal of Medicine* 320 (1989): 427–29.

———. "The Evolution of Therapeutic Research in Cancer." *New England Journal of Medicine* 298 (1978): 907–10.

———. Letter to the Editor. *New England Journal of Medicine* 320 (1989): 472.

Dickens, Charles. *Bleak House.* New York: Norton, 1977.

Dickson, David. *The New Politics of Science.* Chicago: University of Chicago Press, 1988.

Early Breast Cancer Trialists' Collaborative Group (EBCT). "Effects of Adjuvant Tamoxofen and of Cytoxic Therapy on Mortality in Early Breast Cancer: An Overview of 61 Randomized Trials among 28,869 Women." *New England Journal of Medicine* 319 (1988): 1681–92.

Epstein, Steven. *Impure Science: AIDS, Activism and the Politics of Knowledge.* Berkeley and Los Angeles: University of California Press, 1996.

Evans, John. *Playing God: Human Genetic Engineering and the Rationalization of Public Bioethics.* Chicago: University of Chicago Press, 2000.

Ewing, J. "Tissue Reactions to Radiation." *American Journal of Roentgenology and Radiation Therapy* 15 (1926): 93–115.

Faden, Ruth. "Chair's Perspective on the Work of the Advisory Committee on Human Radiation Experiments." *Kennedy Institute of Ethics Journal* 6, no. 3 (1996): 215–21.

Faden, Ruth R., and Tom L. Beauchamp. *A History and Theory of Informed Consent.* New York: Oxford University Press, 1986.

Fisher, Bernard. "Clinical Trials for Cancer." *Cancer* 54 (1984): 2609–17.

———. "Laboratory and Clinical Research in Breast Cancer—a Personal Adventure: The David A. Karnofsky Memorial Lecture." *Cancer Research* 40 (1980): 3863–74.

Fisher, Bernard, Andrew Glass, Carol Redmond, et al. "L-Phenylalanine Mustard (L-PAM) in the Management of Primary Breast Cancer: An Update of Earlier Findings and a Comparison with Those Utilizing L-PAM Plus 5-Fluorouracil (5-FU)." *Cancer* 39 (1977): 2883–2903.

Fleck, Ludwik. *Genesis and Development of a Scientific Fact.* 1935. Edited by Thaddeus J. Trenn and Robert K. Merton. Translated by Fred Bradley and Thaddeus J. Trenn. Chicago: University of Chicago Press, 1979.

———. "The Problem of the Science of Science" (1946). Reprinted in *Cognition and Fact: Materials on Ludwik Fleck,* ed. Robert S. Cohen and Thomas Schnelle. Dordrecht: D. Reidel, 1986.

Forrester, John. "If p, Then What? Thinking in Cases." *History of the Human Sciences* 91 (1996): 1–25.

Foucault, Michel. "Governmentality." In *The Foucault Effect: Studies in Governmentality,* ed. Graham Burchell, Colin Gordon, and Peter Miller. Chicago: University of Chicago Press, 1991.

Fox, Renee. *The Sociology of Medicine: A Participant Observer's View.* Englewood Cliffs, NJ: Prentice-Hall, 1989.

Frankel, Mark S. "The Development of Policy Guidelines Governing Human Experimentation in the United States: A Case Study of Public Policy-Making for Science and Technology." *Ethics in Science and Medicine* 2 (1975): 43–59.

Fredrickson, Donald S. "Seeking Technical Consensus on Medical Interventions." *Clinical Research* 26 (1978): 116–17.

Freedman, Benjamin. "Equipoise and the Ethics of Clinical Research." *New England Journal of Medicine* 317 (1987): 141–45.

Freireich, Emil J., and Noreen A. Lemak. *Milestones in Leukemia Research.* Baltimore: Johns Hopkins University Press, 1991.

Fried, Charles. *Medical Experimentation: Personal Integrity and Social Policy.* Amsterdam: North-Holland, 1974.

Galison Peter. 1999. "Trading Zone: Coordinating Action and Belief." In *Science Studies Reader,* ed. Mario Biagioli. New York: Routledge, 1999.

Gehan, Edmund A., and Noreen A. Lemak. *Statistics in Medical Research: Developments in Clinical Trials.* New York: Plenum Medical, 1994.

Gelber, R. D., and A. Goldhirsch. 1993. "From the Overview to the Patient: How to Interpret Meta-Analysis Data." In *Adjuvant Therapy of Breast Cancer IV* (Recent

Results in Cancer Research, vol. 27), ed. H.-J. Senn, R. D. Gelber, A. Goldhirsch, and B. Thurlman. Berlin: Springer, 1993.

Gieryn, Thomas. *Cultural Boundaries of Science: Credibility on the Line*. Chicago: University of Chicago Press, 1999.

Geison, Gerald L. *The Private Science of Louis Pasteur*. Princeton, NJ: Princeton University Press, 1995.

Ginzburg, Carlo. "Clues: Roots of an Evidential Paradigm." In *Clues, Myths, and the Historical Method*. Baltimore: Johns Hopkins University Press, 1989.

Glick, J. H., R. D. Gelber, A. Goldhirsch, and H.-J. Senn. "Adjuvant Therapy of Primary Breast Cancer: Closing Summary." In *Adjuvant Therapy of Breast Cancer IV* (Recent Results in Cancer Research, vol. 27), ed. H.-J. Senn, R. D. Gelber, A. Goldhirsch, and B. Thurlman. Berlin: Springer, 1993.

Gofman, John. *Radiation and Human Health*. San Francisco: Sierra Club Books, 1981.

Goodman, Jordan, Anthony McElligott, and Lara Marks, eds. *Useful Bodies: Humans in the Service of Medical Science in the Twentieth Century*. Baltimore: Johns Hopkins University Press, 2003.

Gordon, Colin. "Governmentality: An Introduction." In *The Foucault Effect: Studies in Governmentality*, ed. Graham Burchell, Colin Gordon, and Peter Miller. Chicago: University of Chicago Press, 1991.

Greenberg, Daniel S. *The Politics of Pure Science*. New ed. Chicago: University of Chicago Press, 1999.

Grodin, Michael A., ed. *Meta Medical Ethics: The Philosophical Foundations of Bioethics*. Dordrecht: Kluwer Academic, 1995.

Guston, David H. *Between Politics and Science: Assuring the Integrity and Productivity of Research*. Cambridge: Cambridge University Press, 2000.

Guttentag, Otto. "The Problem of Experimentation in Human Beings: The Physician's Point of View." *Science* 117 (1953): 207–10.

Harkness, Jon. "The Significance of the Nuremberg Code." *New England Journal of Medicine* 338, no. 14 (1998): 995–96.

Hellman, Samuel, and Deborah S. Hellman. "Of Mice but Not Men: Problems of the Randomized Clinical Trial." *New England Journal of Medicine* 324 (1991): 1565–89.

Hill, A. Bradford. "The Clinical Trial." *British Medical Bulletin* 7 (1951): 278–82.

———. "The Clinical Trial." *New England Journal of Medicine* 247 (1952): 113–19.

Hindess, Barry. *Discourses of Power: From Hobbes to Foucault*. Oxford: Blackwell, 1996.

Hornblum, Alan. *Acres of Skin: Human Experimentation at Holmsburg Prison: A True Story of Abuse and Exploitation in the Name of Medical Science*. London: Routledge, 1998.

Howard-Jones, N. "Human Experimentation and Medical Ethics." In *Proceedings of the XVth CIOMS Round Table Conference, Manila, 13–16 September, 1981*, ed. Z. Bankowski and N. Howard-Jones. Geneva: Council for International Organizations of Medical Sciences, 1982.

Hughes, Thomas. *American Genesis: A Century of Invention and Technological Enthusiasm, 1870–1970*. New York: Penguin, 1990.

Illich, Ivan. *Medical Nemesis: The Expropriation of Health*. London: Calder & Boyars, 1976.

Ivy, Andrew, C. "Nazi War Crimes of a Medical Nature" (1947). Reprinted in *Ethics in Medicine: Historical Perspectives and Contemporary Concerns*, ed. Stanley Joel Reiser, Arthur J. Dyck, and William J. Curran. Cambridge, MA: MIT Press, 1977.

Jacobson, Leon O., R. S. Stone, and J. Garrott. "Physicians and Atomic War." *Journal of the American Medical Association* 139, no. 3 (1949): 138–40.

Jacoby, Itzak. "The Consensus Development Program of the NIH: Current Practice and Historical Perspectives." *International Journal of Technology Assessment in Health Care* 1 (1985): 420–32.

Jasanoff, Sheila. "Contested Boundaries in Policy Relevant Science." *Social Studies of Science* 17 (1987): 195–230.

———. *Designs on Nature: Science and Democracy in Europe and the United States*. Princeton, NJ: Princeton University Press, 2005.

———. *Science at the Bar: Law, Science, and Technology in America*. Cambridge, MA: Harvard University Press, 1995.

———. "Update on Assessment Activities." *International Journal of Technology Assessment in Health Care* 4 (1988): 95–105.

Johns, Harold, and Jack Cunningham. *The Physics of Radiology*. 4th ed. Springfield, IL: Charles Thomas, 1983.

Jonas, Hans. "Philosophical Reflections on Experimenting with Human Subjects." In *Experimentation with Human Subjects*, ed. Paul A. Freund. London: George Allen & Unwin, 1972.

Jonsen, Albert R. *The Birth of Bioethics*. New York: Oxford University Press, 1998.

———. *A Short History of Medical Ethics*. Oxford: Oxford University Press, 2000.

Jonsen, Albert R., and Stephen Toulmin. *The Abuse of Casuistry: A History of Moral Reasoning*. Berkeley and Los Angeles: University of California Press, 1988.

Katz, Jay. "The Consent Principle and the Nuremberg Code: Its Significance Then and Now." In *The Nazi Doctors and the Nuremberg Code: Human Rights in Human Experimentation*, ed. George Annas and Michael A. Grodin. New York: Oxford University Press, 1992.

Keating, Peter, and Alberto Cambrosio. "From Screening to Clinical Research: The Cure of Leukemia and the Early Development of the Cooperative Oncology Groups, 1955–1966." *Bulletin of the History of Medicine* 76 (2002): 299–334.

Kereiakes, James G., William Van de Riet, Clifford Born, et al. "Active Bone-Marrow Dose Related Hematological Changes in Whole-Body and Partial Body Co-60 Gamma Radiation Exposures." *Radiology* 103 (1972): 651–56.

Kevles, Daniel. *The Baltimore Case: A Trial of Politics, Science and Character*. New York: Norton, 1998.

King, Nancy M. P., Gail E. Henderson, and Jane Stein, eds. *Beyond Regulations: Ethics in Human Subjects Research*. Chapel Hill: University of North Carolina Press, 1999.

Kornfeld, Howard. "Nuclear Weapons and Civil Defense: The Influence of the Medical Profession in 1955 and 1983." *Western Journal of Medicine* 38, no. 2 (1983): 207–12.

Kramer, S. "Radiation Therapy Oncology Group and Eastern Cooperative Group Study of Patients with Malignant Gliomas." *Monographs of the National Cancer Institute* 46 (1977): 237–38.

Kula, Witold. *Measures and Men*. Princeton, NJ: Princeton University Press, 1986.

Kutcher, Gerald. "Cancer Clinical Trials and the Transfer of Medical Knowledge: Metrology, Contestation and Local Practice." In *Devices and Designs: Medical Innovations in Historical Perspective*, ed. Carsten Timmermans and Julie Anderson. Houndmills: Palgrave, 2006.

———. Review of *Useful Bodies: Humans in the Service of Medical Science in the Twentieth Century*, ed. Jordan Goodman, Anthony McElligott, and Lara Marks. *History and Philosophy of the Life Sciences* 27, nos. 3–4 (2005): 535.

Ladimer, Irving, and Roger W. Newman, eds. *Clinical Investigation in Medicine: Legal, Ethical and Moral Aspects*. Boston: Law-Medicine Research Institute, 1963.

Lasagna, L. "The Controlled Clinical Trial: Theory and Practice." *Journal of Chronic Diseases* 1 (1955): 353–67.

———. *The Doctors' Dilemmas*. New York: Harper & Bros., 1962.

Lash, Chrisopher. *The True and Only Heaven: Progress and Its Critics*. New York: Norton, 1991.

Laszlo, John. *The Cure of Childhood Leukemia: Into the Age of Miracles*. New Brunswick, NJ: Rutgers University Press, 1995.

Latour, Bruno. *The Pasteurization of France*. Cambridge, MA: Harvard University Press, 1988.

———. *Science in Action: How to Follow Scientists and Engineers through Society*. Cambridge, MA: Harvard University Press, 1987.

Lawrence, Walter, Jr. "Introduction to the Workshop on Clinical Trials." *Cancer* 65 (1990): 2369–70.

———. "Some Problems with Clinical Trials." *Archives of Surgery* 126 (1991): 370–78.

Lederer, Susan. *Subjected to Science: Human Experimentation in America before the Second World War*. Baltimore: Johns Hopkins University Press, 1995.

Legal Environment. "Report on the National Conference on the Legal Environment of Medical Science." In *Clinical Investigation in Medicine: Legal, Ethical and Moral Aspects*, ed. Irving Ladimer and Roger W. Newman. Boston: Law-Medicine Research Institute, 1963.

Lerner, Barron. *The Breast Cancer Wars: Hope, Fear, and the Pursuit of a Cure in Twentieth-Century America*. Oxford: Oxford University Press, 2003.

Leslie, Stuart. *The Cold War and American Science: The Military-Industrial-Academic Complex at MIT and Stanford*. New York: Columbia University Press, 1993.

Lindee, M. Susan. *Suffering Made Real: American Science and the Survivors of Hiroshima*. Chicago: University of Chicago Press, 1994.

Little, Richard N., Jr. "Experimentation with Human Subjects: Legal and Moral Considerations regarding Radiation Treatment of Cancer at the University of Cincinnati College of Medicine." *Atomic Energy Law Journal* 13 (1972): 305–30.

Low-Beer, B. V. A., and R. S. Stone. "Hematological Studies on Patients Treated by Total-Body Exposure of X-Rays." In *Industrial Medicine on the Plutonium Project: Survey and Collected Papers*, ed. Robert S. Stone. New York: McGraw-Hill, 1951.

Löwy, Ilana. *Between Bench and Bedside: Science, Healing and Interleuken-2*. Cambridge, MA: Harvard University Press, 1996.

———. "The Experimental Body." In *Medicine in the Twentieth Century*, ed. Roger Cooter and John Pickstone. Amsterdam: Harwood Academic, 2000.

Ludmerer, Kenneth M. *The Development of American Medical Education*. New York: Basic/Pantheon, 1985.

MacIntyre, Alasdair. *After Virtue: A Study in Moral Theory*. 2nd ed. Guildford: Duckworth, 1985.

Macklin, Ruth. "Disagreement, Consensus and Moral Integrity." *Kennedy Institute of Ethics Journal* 6, no. 3 (1996): 289–311

Marks, Harry M. *The Progress of Experiment: Science and Therapeutic Reform in the United States, 1900–1990*. Cambridge: Cambridge University Press, 1997.

Mathé, G. "Secondary Syndrome: A Stumbling Block in the Treatment of Leukemia by Whole-Body Irradiation and Transfusion of Allogenic Hematopoietic Cells." In *Diagnosis and Treatment of Acute Radiation Injury*. Geneva: WHO, 1961.

McFarland, William, and Howard A. Pearson. "Hematological Events as Dosimeters in Human Total-Body Irradiation." *Radiology* 80 (1963): 850–55.

McGuire, W. L. "Adjuvant Therapy of Node-Negative Breast Cancer." *New England Journal of Medicine* 320 (1989): 425–27.

McKenny, Gerald P. *To Relieve the Human Condition: Bioethics, Technology and the Body*. Albany: State University of New York Press, 1997.

Medical Research Council. "Clinical Trials of Antihistaminic Drugs in the Prevention and Treatment of the Common Cold." *British Medical Journal* 2 (1950): 425–29.

———. "Streptomycin Treatment of Pulmonary Tuberculosis." *British Medical Journal* 2 (1948): 769–82.

Medinger, Fred G., and Lloyd F. Craver. "Total Body Irradiation with Review of Cases." *Radiology* 48 (1942): 651–57.

Meinert, Curtis L. *Clinical Trials: Design, Conduct and Analysis*. Oxford: Oxford University Press, 1986.

Mendelsohn, Everett. "The Politics of Pessimism: Science and Technology circa 1968." In *Technology, Pessimism and Postmodernism*, ed. Yaron Ezrahi, Everett Mendelsohn, and Howard Segal. Dordrecht: Kluwer Academic, 1994.

Merton, Robert K. "The Normative Structure of Science." In *The Sociology of Science: Theoretical and Empirical Investigations* (1942), ed. Norman Storer. Chicago: University of Chicago Press, 1973.

Mills, C. Wright. *The Power Elite*. 1956. New York: Oxford University Press, 1959.

Moreno, Jonathan. *Undue Risk: Secret State Experiments in Humans*. London: Routledge, 2001.

Murtaugh, Joseph S. "Biomedical Sciences." In *Science and the Evolution of Public Policy*, ed. James A. Shannon. New York: Rockefeller University Press, 1973.

National Institutes of Health (NIH) Consensus Conference. "Adjuvant Chemotherapy for Breast Cancer." *Journal of the American Medical Association* 254 (1985): 3461–63.

———. "Treatment of Early Stage Breast Cancer." *Journal of the American Medical Association* 265 (1991): 391–94.

Nickson, J. J. 1951. "Blood Changes in Human Beings Following Total-Body Irradiation." In *Industrial Medicine on the Plutonium Project: Survey and Collected Papers*, ed. Robert S. Stone. New York: McGraw-Hill, 1951.

Nietzsche, Friedrich. *On the Genealogy of Morals*. Translated by Walter Kaufmann and R. J. Hollingwood. 1967. New York: Vintage, 1969.

Nuremberg Code. *Trials of War Criminals before the Nuremberg Military Tribunals under Control Council Law No. 10, Vol. 2*. Washington, DC: U.S. Government Printing Office, 1949.

Omura, George A. "Clinical Cancer Alerts: Less Than Wise." In *Introducing New Treatments for Cancer: Practical, Ethical and Legal Problems*, ed. C. J. Williams. Chichester: John Wiley, 1992.

Osborne, Thomas. "On Liberalism, Neo-Liberalism, and the 'Liberal Profession' of Medicine." *Economy and Society* 3 (1993): 345–56.

Oughterson, Ashley W., and Shields Warren. *Medical Effects of the Atomic Bomb on Japan*. New York: McGraw-Hill, 1956.

Patterson, James T. *The Dread Disease: Cancer and Modern American Culture*. Cambridge, MA: Harvard University Press, 1987.

Patterson, Ralston. "Clinical Trials in Malignant Disease." In *Controlled Clinical Trials: A Conference Organized by the Council for Organizations of Medical Sciences*. Vienna, 1959.

Perley, Sharon, Sev S. Fluss, Zbigniew Bankowski, and Francoise Simon. "The Nuremberg Code: An International Overview." In *The Nazi Doctors and the Nuremberg Code: Human Rights in Human Experimentation*, ed. George Annas and Michael Grodin. New York: Oxford University Press, 1992.

Perry, Seymour, and John T. Kaberer. "The NIH Consensus-Development Program and the Assessment of Health-Care Technologies." *New England Journal of Medicine* 303 (1980): 169–72.

Pickering, Andrew. *The Mangle of Practice: Time, Agency and Science*. Chicago: University of Chicago Press, 1995.

Pocock, Stuart J. *Clinical Trials: A Practical Approach*. Chichester: John Wiley, 1983.

Porter, Theodore M. *Trust in Numbers: The Pursuit of Objectivity in Science and Public Life*. Princeton, NJ: Princeton University Press, 1995.

Prosnitz, L. R. "Medical and Legal Implications of Clinical Alert." *Journal of the National Cancer Institute* 80 (1988): 1574.

Rapoport, Roger. *The Great American Bomb Machine*. New York: Dutton, 1971.

Regan, Louis J. *Doctor and Patient and the Law*. 2nd ed. St. Louis: Mosby, 1949.

―――. *Doctor and Patient and the Law*. 3rd ed. St. Louis: Mosby, 1956.

Reiser, Stanley Joel, Arthur J. Dyck, and William J. Curran, eds. *Ethics in Medicine: Historical Perspectives and Contemporary Concerns*. Cambridge, MA: MIT Press, 1977.

Rettig, Richard A. *Cancer Crusade: The Story of the National Cancer Act of 1971*. Princeton, NJ: Princeton University Press, 1977.

"Review of Mortality Results in Randomised Trials in Early Breast Cancer." *Lancet* 2, no. 8413 (1984): 1205.

Risse, Gunter B. *Mending Bodies, Saving Souls: A History of Hospitals*. New York: Oxford University Press, 1999.

Rose, Nikolas. *Powers of Freedom: Reframing Political Thought*. Cambridge: Cambridge University Press, 1999.

Rose, Nikolas, and Peter Miller. "Political Power beyond the State: Problematics of Government." *British Journal of Sociology* 43 (1992): 173–205.

Rosenberg, Charles E. *The Care of Strangers: The Rise of America's Hospital System*. New York: Basic, 1987.

Rothman, David J. "Ethics and Human Experimentation: Henry Beecher Revisited." *New England Journal of Medicine* 317, no. 19 (1987): 1195–98.

―――. *Strangers at the Bedside: A History of How Law and Bioethics Transformed Medical Decision Making*. New York: Basic, 1991.

Saenger, Eugene. "Effects of Total- and Partial-Body Therapeutic Irradiation in Man." In *Proceedings of the First International Symposium on the Biological Interpretation of Dose from Accelerator-Produced Radiation, Held at the Lawrence Radiation Laboratory, Berkeley, California, March 13–16, 1967* (CONF-670305), ed. Roger Wallace. Washington, DC: U.S. Atomic Energy Commission, Division of Operational Safety, 1967.

―――, ed. *Medical Aspects of Radiation Accidents: A Handbook for Physicians, Health Physicists and Industrial Hygienists*. Washington, DC: Atomic Energy Commission, 1963.

Saenger, Eugene, Edward B. Silberstein, Bernard Aron, et al. "Whole Body and Partial Body Radiotherapy of Advanced Cancer." *American Journal of Roentgenology and Radiation Therapy* 117, no. 3 (1973): 670–85.

Sanderson, S. S. "Irradiation of the Entire Body by the Roentgen Ray: Preliminary Report of 22 Cases." *American Journal of Roentgenology and Radiation Therapy* 35 (1936): 670–80.

Schnelle, Thomas. "Microbiology and Philosophy of Science, Lwow and the German Holocaust: Stations of a Life—Ludwig Fleck, 1896–1961." In *Cognition and Fact: Materials on Ludwig Fleck*, ed. Robert S. Cohen and Thomas Schnelle. Dordrecht: D. Reidel, 1986.

Secretary World Medical Association (WMA). "Editorial." *World Medical Journal* 6 (1960): 95.

Shapin, Steven. "Science and the Public." In *Companion to the History of Modern Science*, ed. R. C. Olby, G. N. Cantor, J. R. R. Christie, and M. J. S. Hodge. London: Routledge, 1990.

————. *A Social History of Truth: Civility and Science in Seventeenth Century England.* Chicago: University of Chicago Press, 1994.

Shapin, Steven, and Simon Schaffer. *Leviathan and the Airpump: Hobbes, Boyle and the Experimental Life.* Princeton, NJ: Princeton University Press, 1985.

Shapiro, Arthur K., and Elaine Shapiro. *The Powerful Placebo: From Ancient Priest to Modern Physician.* Baltimore: Johns Hopkins University Press, 1997.

Sharpe, Virginia A., and Alan I. Faden. *Medical Harm: Historical, Conceptual, and Ethical Dimensions of Iatrogenic Illness.* Oxford: Oxford University Press, 1998.

Shevell, Michael I. "Neurology's Witness to History: The Combined Intelligence Operative Sub-Committee Reports of Leo Alexander." *Neurology* 47 (1996): 1096–1103.

Shimkin, Michael. "The Problem of Experimentation on Human Beings: The Research Worker's Point of View." *Science* 117 (1953): 205–7.

Shuster, Evelyne. "Fifty Years Later: The Significance of the Nuremberg Code." *New England Journal of Medicine* 337, no. 20 (1997): 1436–40.

Singer, Peter, ed. *A Companion to Bioethics.* Oxford: Basil Blackwell, 1991.

Sluys, D. Félix. "La roentgenisation totale dans la lymphogranulomatose maligne." *Journal belge de radiologie* 19 (1930): 297–317.

Star, Susan Leigh, and James R. Griesemer. "Institution, Ecology, 'Translation,' and Boundary Objects: Amateurs and Professionals in Berkeley's Museum of Vertebrate Zoology." *Social Studies of Science* 19 (1989): 387–420.

Starr, Paul. *The Transformation of American Medicine.* New York: Basic, 1982.

Stephens, Jerome. "Political, Social, and Scientific Aspects of Medical Research on Humans." *Politics and Society* 3, no. 4 (1973): 409–35.

Stephens, Martha. *The Treatment: The Story of Those Who Died in the Cincinnati Radiation Tests.* Durham, NC: Duke University Press, 2002.

Stevens, M. L. Tina. *Bioethics in America: Origins and Cultural Politics.* Baltimore: Johns Hopkins University Press, 2000.

Stevens, Rosemary. *In Sickness and in Wealth: American Hospitals in the Twentieth Century.* New York: Basic, 1989.

Stone, Robert S., ed. *Industrial Medicine on the Plutonium Project: Survey and Collected Papers.* New York: McGraw-Hill, 1951.

Strickland, Steven. *Politics, Science and Dread Disease: A Short History of United States Medical Research Policy.* Cambridge, MA: Harvard University Press, 1972.

Sullivan, Mark D. "Placebo Controls and Epistemic Control in Orthodox Medicine." *Journal of Medicine and Philosophy* 18 (1993): 213–31.

Task Force of the Presidential Advisory Group. "The Science Court Experiment: An Interim Report." *Science* 20 (1976): 653–56.

Taylor, Charles A. *Defining Science: A Rhetoric of Demarcation.* Madison: University of Wisconsin Press, 1996.

Teschendorff, W. "Ueber Bestrahlung des ganzen menschlichen Korpes bei Blutkrankheiten." *Strahlentherapie* 26 (1927): 720–28.

Thomas, E. Donnal. "Observations in Supralethal Whole Body Irradiation and Marrow Transplantation in Man and Dog." In "Physical Factors and Modification of Radiation Injury," ed. L. D. Hamilton. Special issue, *Annals of the New York Academy of Sciences* 114 (1964): 393–96.

Thomas, E. Donnal, C. Dean Buckner, Meera Banaji, et al. "One Hundred Patients with Acute Leukemia Treated by Chemotherapy, Total Body Irradiation, and Allogeneic Marrow Transplantation." *Blood* 49, no. 4 (1977): 511–33.

Thomas, E. Donnal, Harry L. Locte Jr., and Joseph W. Ferrebee. "Irradiation of the Entire Body and Marrow Transplantation: Some Observations and Comments." *Blood* 14 (1959): 1–23.

Thomas, E. Donnal, Harry L. Locte Jr., Wang Ching Lu, and Joseph W. Ferrebee. "Intravenous Infusion of Bone Marrow in Patients Receiving Radiation and Chemotherapy." *New England Journal of Medicine* 257 (1957): 491–96.

Thomas, E. Donnal, Rainer Storb, and Reginal A. Clift. "Bone Marrow Transplantation." Pt. 1. *New England Journal of Medicine* 292 (1975): 832–43.

Thompson, E. P. "The Moral Economy of the Crowd." In *Customs in Common*. Harmondsworth: Penguin, 1991.

Trentin, John J. "Consequences of Bone Marrow Induced Radiation Recovery." In *Radiation Biology and Cancer: Collection of Papers Presented at the Twelfth Annual Symposium on Fundamental Cancer Research, 1958*. London: Peter Owen, 1960.

University of Texas. *The First Twenty Years of the University of Texas M. D. Anderson Hospital and Tumor Institute*. Houston: Printing Division of the University of Texas, 1964.

U.S. Department of Health, Education and Welfare (USDHEW). Office for Protection of Risks. *Belmont Report: Ethical Principles and Guidelines for the Protection of Human Subjects of Research*. Washington, DC: U.S. Government Printing Office, 1979.

Warren, Shields. "The Pathological Effects of an Instantaneous Dose of Radiation." *Cancer Research* 6 (1946): 449–53.

Welsome, Eileen. *The Plutonium Files: America's Secret Medical Experiments in the Cold War*. New York: Dial, 1999.

Welt, Louis G. "Reflections on the Problems of Human Experimentation" (1961). Reprinted in *Clinical Investigation in Medicine: Legal, Ethical and Moral Aspects*, ed. Irving Ladimer and Roger W. Newman. Boston: Law-Medicine Research Institute, 1963.

Whittemore, Gilbert. "A Crystal Ball in the Shadows of Nuremberg and Hiroshima: The Ethical Debate over Human Experimentation to Develop a Nuclear-Powered Bomber, 1946–51." In *Science, Technology and the Military*, ed. Everett Mendelsohn, Merritt Roe Smith, and Peter Weingart. Dordrecht: Kluwer, 1988.

Wiggers, Carl J. "Human Experimentation as Exemplified by the Career of Dr. William Beaumont" (1950). Reprinted in *Clinical Investigation in Medicine: Legal, Ethical and Moral Aspects*, ed. Irving Ladimer and Roger W. Newman. Boston: Law-Medicine Research Institute, 1963.

Winkler, Allen M. *Life under a Cloud: American Anxiety about the Atom.* New York: Oxford University Press, 1993.

Winslade, William J., and Todd L. Krause. "The Nuremberg Code Turns Fifty." In *Ethics Codes in Medicine: Foundations and Achievements of Codification since 1947,* ed. Ulrich Trohler and Stella Reiter-Theil. Aldershot: Ashgate, 1998.

Zubrod, C. Gordon. "Clinical Trials in Cancer Patients: An Introduction." *Controlled Clinical Trials* 3 (1982): 185–87.

———. "Origins and Development of Chemotherapy Research at the National Cancer Institute." *Cancer Treatment Reports* 68 (1984): 9–19.

Zubrod, C. Gordon, Saul A. Schepartz, and Stephen K. Carter. "Historical Background of the National Cancer Institute's Drug Development Thrust." *National Cancer Institute Monographs* 45 (1972): 7–11.

INDEX